Cancer Therapy

Differentiation, Immunomodulation and Angiogenesis

NATO ASI Series

Advanced Science Institutes Series

A series presenting the results of activities sponsored by the NATO Science Committee, which aims at the dissemination of advanced scientific and technological knowledge, with a view to strengthening links between scientific communities.

The Series is published by an international board of publishers in conjunction with the NATO Scientific Affairs Division

A Life Sciences B Physics	Plenum Publishing Corporation London and New York
C Mathematical and Physical Sciences D Behavioural and Social Sciences E Applied Sciences	Kluwer Academic Publishers Dordrecht, Boston and London
F Computer and Systems Sciences G Ecological Sciences H Cell Biology I Global Environmental Change	Springer-Verlag Berlin Heidelberg New York London Paris Tokyo Hong Kong Barcelona Budapest

NATO-PCO DATABASE

The electronic index to the NATO ASI Series provides full bibliographical references (with keywords and/or abstracts) to more than 30000 contributions from international scientists published in all sections of the NATO ASI Series. Access to the NATO-PCO DATABASE compiled by the NATO Publication Coordination Office is possible in two ways:

- via online FILE 128 (NATO-PCO DATABASE) hosted by ESRIN, Via Galileo Galilei, I-00044 Frascati, Italy.
- via CD-ROM "NATO Science & Technology Disk" with user-friendly retrieval software in English, French and German (© WTV GmbH and DATAWARE Technologies Inc. 1992).

The CD-ROM can be ordered through any member of the Board of Publishers or through NATO-PCO, Overijse, Belgium.

Series H: Cell Biology, Vol. 75

Cancer Therapy

Differentiation, Immunomodulation and Angiogenesis

Edited by

Natale D'Alessandro

Institute of Pharmacology
Piazza XX Settembre 4, 98100 Messina, Italy

Enrico Mihich

Grace Cancer Drug Center
Department of Experimental Therapeutics
Roswell Park Center Institute
Elm and Carlton Streets, Buffalo, NY 14263, USA

Luciano Rausa

Faculty of Medicine
Policlinic of the University "P. Giaccone"
Via del Vespro 129, 90127 Palermo, Italy

Haim Tapiero

Laboratoire de Pharmacologie Cellulaire & Moleculaire
ICIG, Hopital Paul-Brousse
14 av. Paul-Vaillant-Couturier, 94800 Villejuif, France

Thomas R. Tritton

University of Vermont, School of Medicine
Burlington, VT 05405, USA

Springer-Verlag
Berlin Heidelberg New York London Paris Tokyo
Hong Kong Barcelona Budapest
Published in cooperation with NATO Scientific Affairs Division

Proceedings of the NATO Advanced Study Institute on Specific Approaches in Cancer Therapy: Differentiation, Immunomodulation and Angiogenesis held at Erice, Italy, October 17–27, 1992

ISBN-13:978-3-642-84615-1 e-ISBN-13:978-3-642-84613-7
DOI: 10.1007/978-3-642-84613-7

Softcover reprint of the hardcover 1st edition 1993

Typesetting: Camera ready by authors
31/3145 - 5 4 3 2 1 0 - Printed on acid-free paper

FOREWORD

The insufficient selectivity of antitumour drugs currently available and the frequent phenomenon of drug resistance represent a major obstacle to further advances in cancer treatment. With this in mind, the aim of the NATO ASI held from 17 to 27 october 1992 at the "Ettore Majorana Centre for Scientific Culture" in Erice, of which this book represents the proceedings, was to examine comprehensively the basic and clinical conditions of some innovative approaches, i.e. differentiation, immunomodulation and inhibition of angiogenesis, which may support or even substitute chemotherapy.

A general consensus was that cancer cells are characterized by being arrested at an immature level of development while retaining their proliferative capacity; a rational approach thus involves the induction of tumour cell differentiation to a mature stage, where proliferation ceases. However, Alain Zweibaum (Villejuif, France) warned that, in colon cancer, a small proportion of cells capable of differentiating possess particular properties which allow them to escape stress conditions and to start again their growth.

A deep insight into the mechanisms underlying the control of cell multiplication and differentiation was provided by various lectures. Eliezer Huberman (Argonne, Illinois) showed that the signal transduction pathway, which mediates phorbol 12-myristate 13-acetate-induced differentiation in the human promyelocytic HL-60 cells, requires specific protein kinase C isozymes (β, a "δ-like") for the proper expression of the early response genes such as *junB*, *c-fos* and *c-jun*. Raymond Frade (Paris, France) analyzed the signal transduction from the Epstein-Barr virus receptor (EBV/C3dR) in human B lymphocytes and demonstrated that, in the human B lymphoma cell line Raji, EBV/C3dR interacts specifically with the p53 cellular tumour suppressor gene- encoded phosphoprotein, which is not expressed in normal B cells; in normal B lymphocytes, EBV/C3dR interacts with p68, an intracellular calcium-binding protein belonging to the annexin VI family. Thomas R. Tritton (Burlington, Vermont) indicated that the ability of anticancer drugs like doxorubicin to kill susceptible cells involves a series of events initiated at the plasma membrane and proceeding through the protein kinase C signal transduction pathway to ultimate damage to the DNA in the nucleus. In the colon carcinoma cell lines described by Michael G. Brattain (Toledo, Ohio), a progressed, highly aggressive, phenotype is accompanied by a strong internal TGF-α autocrine loop, which leads to independence from regulatory growth factors. Thus, the possibility raises of developing therapeutic approaches on the inhibition of TGF-α transcription.

Other firm experimental therapeutic options involving differentiation were outlined by Giovanni B. Rossi (Rome, Italy), who discussed the role of interferons in the differentiation of Friend erythroleukemia, and by Alexander Bloch (Buffalo, New York), who pointed out that DNA-specific antitumour agents, at concentrations that are minimally cytotoxic, are capable of increasing the responsiveness of cancer cells to differentiation signals. Finally, Laurent Degos (Paris, France) showed how all- trans retinoic acid induces a very high rate of complete remissions in acute promyelocytic leukemia, which represents indeed the first model of differentiation therapy in human malignancies.

Modulation by biological agents, cytotoxic effector cells and drugs was considered in attempts to boost endogenous antitumour defenses and/or to render neoplastic cells more susceptible to the host attack. Enrico Mihich (Buffalo, New York) presented the influence of doxorubicin on, both specific and natural, immune functions and the curative effects of this drug plus IL-2 or TNF in the EL4 lymphoma in the C57Bl/6 mouse, as a proper example of the useful interaction between an anticancer drug with immunomodulatory activity and certain cytokines with antitumour action. The interrelationship between anticancer drugs and immunity was again underscored by Jean-Luc Teillaud (Paris, France), who pinpointed the different ways doxorubicin, pirarubicin or aclacinomycin act on the B cell system. In addition, Paolo Puccetti (Perugia, Italy) described how potent mutagenic compounds, such as triazene and nitrosoguanidine derivatives, generate highly immunogenic cell variants of murine lymphomas; Benjamin Bonavida (Los Angeles, California), suggested that the combination of TNF-α and either drugs (doxorubicin, CDDP) or diphteria toxin may overcome tumour resistance to either one or both agents.

In the clinical context, Carlo Gambacorti Passerini (Milan, Italy) reviewed the use of IL-2 in advanced cancer, raising the question of whether, besides LAK cells, other subpopulations of lymphocytes could be responsible for the clinical effects of the treatment. Fiorella Guadagni (Rome, Italy) cited the possibility of improving the diagnostic and therapeutic efficacy of monoclonal antibodies by the use of interferons, which can upregulate human tumour antigens.

In the last section, the importance of interfering with tumour blood vessel formation and function, and its potential in the treatment of metastasis, was taken into account. Claudio J. Conti (Smithville, Texas) made a complete report on the biology of angiogenesis, its factors and genetic regulation. Francesco Colotta (Milan, Italy) described the complex, ambiguous role of tumour associated macrophages in the regulation of primary tumour growth, angiogenesis and metastasis. Ralph J. Bernacki (Buffalo, New York) provided evidence of the role played by specific lectins of the extracellular matrix, such as galaptin, in tumour cell adhesion, suggesting the therapeutic exploitation of newly synthesized membrane sugar analogues as modulators of cell surface structure and adhesion.

The Course was attended by about 90 selected participants from Albania, Bulgaria, Canada, Czechoslovakia, France, Germany, Greece, Italy, Moldavia, Poland, Portugal, Romania, Russia, Ukraine, USA and Turkey. The lectures were followed by intense and exhaustive discussions; among the many interesting interventions, there were those, presented as formal papers in this book, of Asterios S. Tsiftosoglou (Thessaloniki, Greece) on differentiation, Nicolò Borsellino (Palermo, Italy), Vasile F. Dima (Bucharest, Romania), Maurizio R. Soma (Milan, Italy) and Francesco Squadrito (Messina, Italy) on immunomodulation, and of Romano Danesi (Pisa, Italy) and Marina Ziche (Florence, Italy) on angiogenesis.

To sum up, the participants in the Erice meeting were exposed to new concepts and/or models to be considered or applied in their laboratories or clinical practice. Indeed, they saw evidence indicating that the exploitation of discrete mechanisms in differentiation, immunomodulation or angiogenesis may be of therapeutic value, at least in selected human neoplasms. This situation could well improve in the next few years.

Finally, the Course was a significant occasion for personal contact, especially between scientists from East Europe and those from the NATO area. We wish to express our appreciation to our main sponsoring Institutions, the NATO Scientific Affairs Division, the Ettore Majorana Centre for Scientific Culture, and the Sicilian Regional Government for their moral and financial support. Last, but not least, we gratefully acknowledge Dr. Carla Flandina for her patience, care and hard work in dealing with the organization of the course.

The editors.

TABLE OF CONTENTS

COLON CANCER CELL DIFFERENTIATION AS RELATED TO METHOTREXATE AND 5-FLUOROURACIL RESISTANCE

A. Zweibaum, T. Lesuffleur, A. Barbat, E. Dussaulx, I. Chantret, L. Mahraoui, G. Chevalier, E. Brot-Laroche, and M. Rousset
INSERM U178, 16 Av. Paul Vaillant Couturier, 94807 Villejuif Cedex, France.

INTRODUCTION

Colon cancer is a major health problem because of its high frequency and the poor outcome of invasive forms, due to their overall resistance to chemotherapy. Although it has been assumed that drug resistance could be associated with some particular cell populations, these have not been characterized yet. It is only in recent years that progress in the field of intestinal cell biology, based on the development of cultured cell lines and availability of immunological and molecular probes, has allowed to characterize cells at the single cell level and to study their organization and functions. This has led to the concept of colon cancer cell differentiation. Cellular differentiation, which should not be confused with differentiation of colon cancers as defined by pathologists, is the ability of colon cancer cells to express, *in vitro* and *in vivo*, the same morphological and functional characteristics as normal epithelial intestinal cells, i.e. enterocytes or goblet cells. Whether these cells, which behave like normal cells as to their differentiation characteristics and functions, possess particular adaptation properties which allow them to escape the cytotoxic effect of a number of stress conditions, including treatment with anti-cancer drugs, is based on recent experimental data obtained with cultured human colon cancer cell lines. The purpose of this article is to (1) summarize our knowledge on colon cancer cell differentiation and show how experience in the field of cell biology can be transferred to clinical situations and (2) focus on experimental data which suggest that drug resistance is associated with cellular differentiation.

NATO ASI Series, Vol. H 75
Cancer Therapy
Edited by N. D'Alessandro, E. Mihich, L. Rausa, H. Tapiero, and T. R. Tritton

DIFFERENTIATION OF COLON CANCER CELL LINES: CACO-2 AND HT-29 CELL LINES AS A PARADIGM FOR COLON CANCER CELL DIFFERENTIATION

The cell lines HT-29 (Fogh & Trempe, 1975) and Caco-2 (Fogh et al., 1977) have been established by Dr. Jorgen Fogh (Memorial Sloan Kettering Cancer Center, NY) in 1964 and 1974 respectively. The first demonstration of their ability to differentiate dates back to 1982 (Pinto et al.) and 1983 (Pinto et al.) for HT-29 and Caco-2 cells respectively. Before any description of the differentiation characteristics associated with these cell lines it is essential to point out that **the differentiation of these cells is growth-related with the onset of the differentiation process taking place after the cells have reached confluency:** exponentially growing cells are never differentiated, whatever their commitment to differentiation; differentiation progressively takes place after confluency and is complete only at late confluency (Fig. 1). Therefore all the descriptions reported below refer to cells at late confluency.

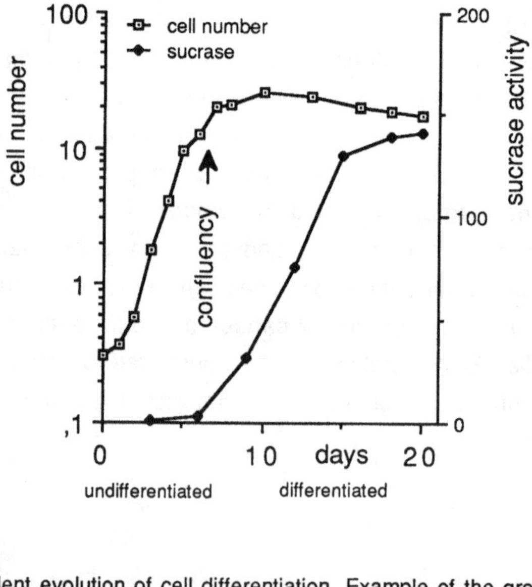

Fig.1. Growth-dependent evolution of cell differentiation. Example of the growth curve of Caco-2 cells and of the growth-related increase in sucrase activity used as a quantitative control of differentiation.

Caco-2 cells: a model for enterocytic differentiation

Caco-2 cells spontaneously express a typical enterocytic differentiation: they form a monolayer of polarized cells with the presence of an apical brush border (Fig.2). The membrane of the brush border microvilli is endowed with the same hydrolases as expressed by the normal small intestine or the fetal colon (Lacroix et al., 1984, Zweibaum et al., 1984), namely sucrase-isomaltase, lactase, aminopeptidase N, dipeptidylpeptidase-IV and alkaline phosphatase (Hauri et al, 1985, Pinto et al, 1983). Their cytoskeleton expresses the Ca^{2+}-binding protein villin (Robine et al., 1985).

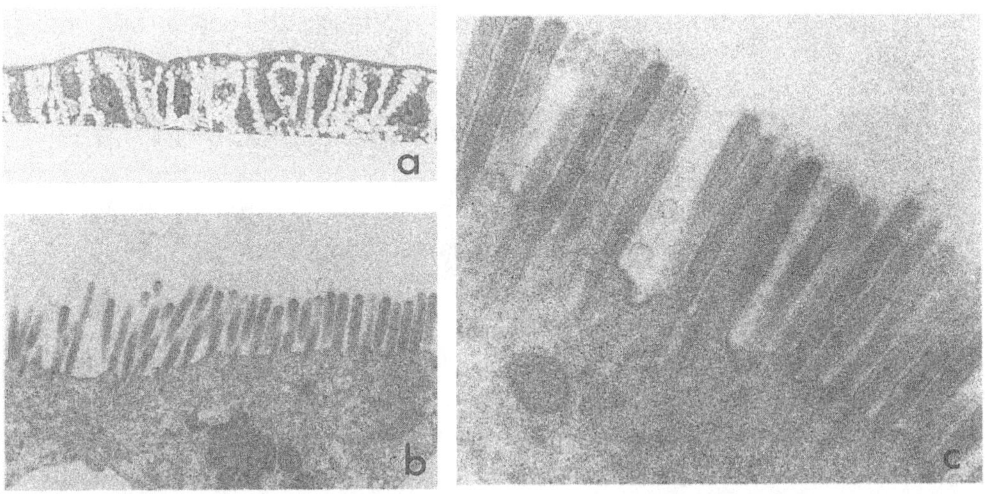

Fig.2. Morphological differentiation of the cell layer of post-confluent Caco-2 cells. (a) thin section of the cell layer perpendicular to the support showing the polarized organization of the cells (x120); (b) detail at transmission electron microscopy of the apical brush border (x12,000) and (c) (x30,000).

Because of these differentiation characteristics they have been extensively used as a model for normal intestinal cells. Recent work in the field includes the polarized delivery of membrane or secreted proteins such as brush border-associated hydrolases or basolateral membrane-associated proteins (Eilers et al., 1989, Gilbert et al., 1991, Hughson & Hopkins, 1990, Le Bivic et al., 1990, Matter et al., 1990, Van't Hoff & Van Meer, 1990) and apolipoproteins (Rindler & Traber, 1988), the genetic control of gene products associated with the functions of intestinal epithelial cells like sucrase-isomaltase (Chantret et al., 1992, Cross & Quaroni, 1991, Rousset et al., 1989), dipeptidylpeptidase-IV (Darmoul et al., 1991), alkaline phosphatase (Matsumoto et al., 1990), carcinoembryonic antigen (Hawk & Stanners, 1991) or apolipoproteins (Daschti et al., 1990, Field et al., 1987, Hughes et al., 1988, Moberly et al., 1990), ion absorption and secretion (Bear & Reyes, 1992, Giuliano & Wood, 1991, Grasset et al, 1984, Sood et al., 1992, Watson et al., 1991,), transport of vitamins such as vitamin B12 (Dix et al., 1990, Ramanujam et al., 1991) and vitamin D (Giuliano et al., 1991), transport of bile acids (Hidalgo & Borchardt, 1990, Woodcock et al., 1991), absorption of sugars (Blais et al., 1991, Mahraoui et al., 1992), absorption, metabolism and secretion of lipids (Faust & Albers, 1988, Field et al., 1988, Trotter & Storch, 1991), drug transport (Audus et al., 1990, Dantzig & Bergin, 1990, Hilgers & Barton, 1988, Wilson et al., 1990), interrelation with the extracellular matrix (Bouziges et al., 1991, Kaiserlian et al., 1991), expression of peptide receptors (Laburthe et al., 1987), interactions with microorganisms of the intestinal ecosystem (Darfeuille-Michaud et al., 1990, Gaillard et al., 1991).

HT-29 cells: a heterogeneous population and a pluripotent model for enterocytic, ion transporting and mucus-secreting cells

In contrast to Caco-2 cells, the HT-29 cell line is heterogeneous and contains only a small proportion of differentiated cell types (<5%) which express three main phenotypes: enterocytes, goblet cells and ion-transporting cells (Lesuffleur et al, 1990, 1991c). Under various pressure conditions, the first of which being the replacement of glucose by galactose in the culture medium (Pinto et al., 1982), a number of subpopulations or clones have been isolated which are either enterocytic (Hafez et al., 1990, Huet et al., 1987, Laboisse et al., 1988, Pinto et al., 1982, Wice et al., 1985, Zweibaum et al., 1985), or mucus-secreting (Fig.3) (Augeron & Laboisse, 1984, Hafez et al., 1990, Huet et al., 1987, Kreusel et al., 1991, Lesuffleur et al., 1990, 1991a, 1991c) or ion-transporting, these latter forming domes (Augeron & Laboisse, 1984, Fantini et al., 1986, Hafez et al., 1990). These differentiated HT-29 populations have also been extensively used as models for the regulation of functions such as Cl^- secretion (Kreusel et al., 1991, Tilly et al., 1991), transport of Ca^{2+} (Fischer et al., 1992) and sugars (Blais, 1991), mucin secretion (Phillips et al., 1988, Dahiya et al., 1992), endocytic processes (Godefroy et al.,

1990), metabolism of low density lipoproteins (Viallard et al., 1990), or polarized expression of membrane proteins such as α2-adrenergic receptors (Devedjian et al., 1991), VIP receptors (Fantini et al., 1988, Laburthe et al., 1978), HLA antigens and transferrin receptors (Godefroy et al., 1988), the product of the CFTR gene (Crawford et al., 1991, Montrose-Rafizadeh et al., 1991, Sood et al., 1992, Zeitlin et al., 1992) or of cytoskeleton-associated proteins like villin (Dudouet et al., 1987, Pringault et al., 1986).

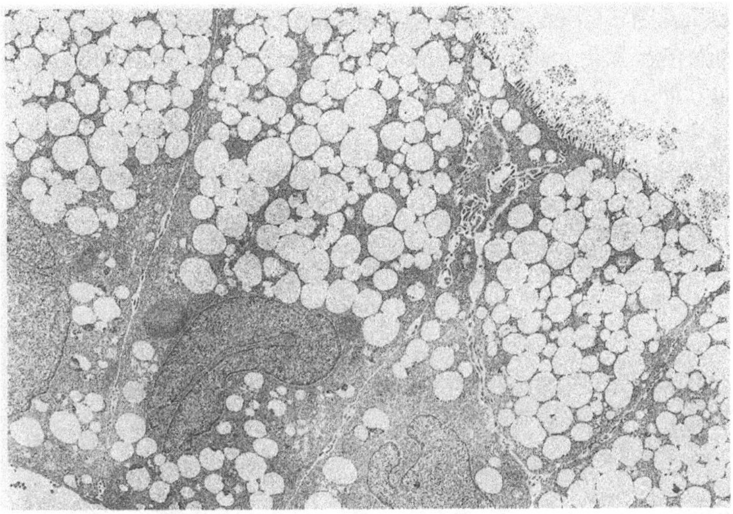

Fig. 3. Electron microscopy micrograph of a section of mucus-secreting HT-29 cells (x10,000).

Altogether the tremendous amount of data accumulated over the past 5 years with these two cell lines leads to the evidence that **some colon cancer cells behave like normal cells** as to their differentiation characteristics and functions.

CACO-2 AND HT-29 CELLS AS MODELS FOR THE IMMUNOHISTOLOGICAL DETECTION OF DIFFERENTIATED CELL TYPES IN CLINICAL SPECIMENS

It is clear that most of the methodologies which allow the characterization of the differentiation-associated features and functions of Caco-2 and HT-29 cells are not adapted to the characterization of similar differentiated cell types in clinical specimens. However it has been shown, in the course of these studies, that a number of antibodies, which have been developed against normal gene products associated with normal intestinal epithelial cells, will react with Caco-2 and HT-29 cells according to patterns which are related to their differentiation status. More precisely antibodies specific for brush border-associated proteins will show, as e.g. in the small intestine (Fig. 4a), an apical expression in sections of the cell layer of these differentiated cultured cells (Fig. 4b). It can be assumed that when a similar staining pattern is encountered in sections of a colonic tumor (Fig. 4c) it is indicative, as confirmed by electron microscopy (Zweibaum et al., 1983), of the presence of cells with a similar type of differentiation.

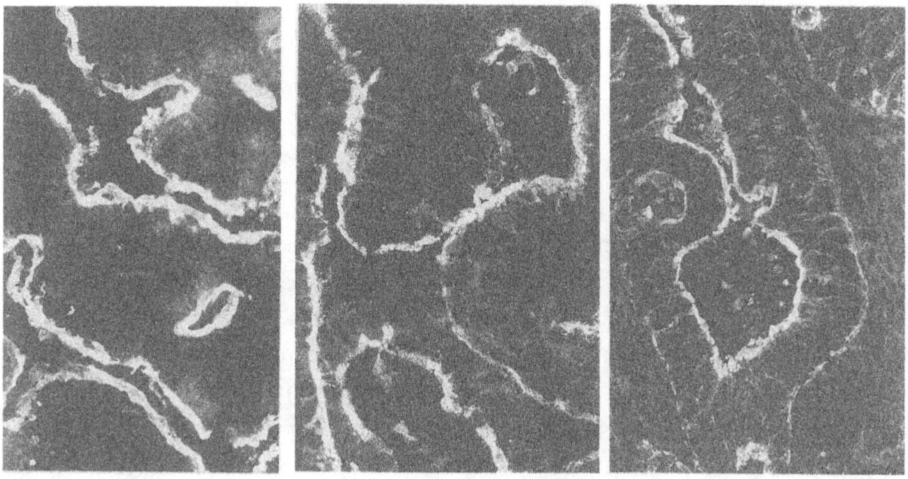

Fig.4. An example of immunofluorescence detection of sucrase-isomaltase associated with (left) the brush border of the normal small intestine, (middle) the brush border of Caco-2 cells (for details on the methodology see Rousset et al., 1989) and (right) apical structures in a colon cancer. Immunofluorescence was performed on cryostat sections (x150)

The few studies which have been performed so far suggest that such differentiated cell types are present , at various proportions, in most colon cancers (Carboni et al., 1987, Czernichow et al., 1989, Moll et al., 1987, Real et al., 1992, Wiltz et al., 1990, Zweibaum et al., 1984). However most of these studies have been restricted to a limited number of tumors and of differentiation-associated markers. Further characterization of which differentiated cell types are present in clinical samples will need the use of a large panel of antibodies specific for all the gene products which are already known to be associated with the differentiation of HT-29 and Caco-2 cells, e.g. not only brush border hydrolases or villin but also CFTR, mucin peptides, hexose transporters, extracellular matrix components and adhesion molecules etc. In the meantime the results obtained in colon cancers already allow to assume that the differentiated cell types found in colon cancers are the offsprings of a particular cell lineage committed to differentiate. What is the biological significance of such cells? A possible answer, here again, relies on experimental data obtained with cultured cell lines.

METABOLIC RESISTANCE AND ADAPTATION OF CELLS COMMITTED TO DIFFERENTIATION.

Evidence that cells with a differentiation potential do have particular properties of resistance and adaptation to "metabolic stress" relies on observations with the HT-29 cell line. As already mentioned, the differentiated populations isolated from this heterogeneous cell line have always been obtained under pressure conditions (for detailed references see Lesuffleur et al., 1991c) such as replacement of glucose by galactose (Huet et al., 1987, Pinto et al., 1982), glucose deprivation (Zweibaum et al., 1985), or treatment with drugs such as sodium butyrate (Augeron & Laboisse, 1984) or HMBA (Hafez et al., 1990). In all these conditions the occurrence of stable differentiated populations is the result of a same sequence of events with an initial phase of high mortality followed by a progressive growth adaptation to these conditions. This growth adaptation is followed by the emergence of differentiated cell populations (Fig. 5). Similar results have been obtained with two other cell lines, namely HCT-GEO and HCT-EB (Brattain et al., 1981) when adapted to glucose deprivation (Chantret et al., 1988). Although the mechanisms involved in metabolic adaptation and differentiation are still unknown it can be postulated that these differentiated populations evolve from this small population of cells that are present in the original cell line and are committed to differentiation.

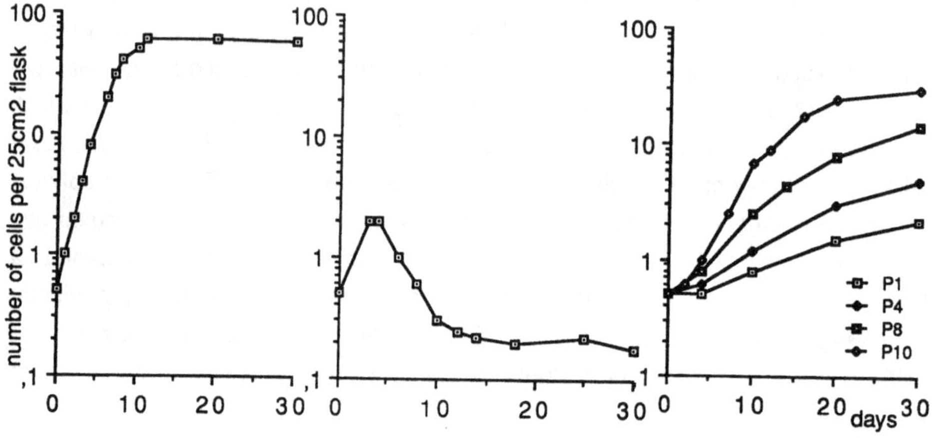

Fig.5. Metabolic adaptation of HT-29 cells to glucose deprivation. Left: growth curve of HT-29 cells cultured in the presence of glucose. Middle: same cells seeded in glucose-free medium showing a high rate of mortality with only a few surviving cells which remain viable after 30 days. Right: passage-related increase of cell growth at serial passages (P) in glucose-free medium

RESISTANCE AND ADAPTATION TO METHOTREXATE AND 5-FLUOROURACIL OF HT-29 CELLS COMMITTED TO DIFFERENTIATION

Whether these properties of metabolic resistance and adaptation of HT-29 cells with a differentiation potential also apply to metabolic anti-cancer drugs such as methotrexate (MTX) and 5-fluorouracil (5-FU) led us to extend to both drugs the same protocols as used for adaptation to glucose deprivation (Zweibaum et al., 1985). HT-29 cells were cultured in the presence of increasing concentrations of both drugs, starting at concentrations immediately superior to the LD_{50} ($3.3x10^{-8}$M for MTX and $0.7x10^{-6}$M for 5-FU). The same sequence of events as with glucose deprivation was observed for each level of drug concentration, i.e. a high rate of mortality followed by a progressive resumption of cell growth in the following passages (See Lesuffleur et al.,1990, 1991a, c). Normally growing populations were obtained which were adapted respectively to 10^{-7}, 10^{-6}, 10^{-5}, 10^{-4} and 10^{-3}M MTX (Lesuffleur et al., 1990, 1991a) and to $1x10^{-6}$, $5x10^{-6}$, $1x10^{-5}$ and 2x10-

^{5}M 5-FU (Lesuffleur et al., 1991c). As with glucose deprivation, adaptation to either drug results in the selection of differentiated populations.

The whole process shares a number of characteristics which should be considered:

(1) The emergence of populations with differentiation potential (i.e. populations which are totally differentiated at late confluency) occurs as soon as the cells are stably adapted to the first supra-LD_{50} concentration of each drug ($1x10^{-7}$M for MTX, $1x10^{-6}$M for 5-FU). Further increase in drug resistance does not modify significantly the differentiation characteristics of the cells, except for the highest concentrations of MTX (10-4 and 10-3M) which show a shift in the differentiation pattern of the cells (Lesuffleur et al., 1991a).

(2) The ability of the cells to differentiate is irreversible. This means that the cells maintain the same differentiation potential when they are switched back to drug-free medium.

(3) The differentiation pattern of the cells differs according to the drug they are adapted to: MTX-adapted cells form an homogeneous population of polarized goblet cells which secrete mucins of gastric immunoreactivity and exhibit, like in the fetal colon, an apical brush border endowed with dipeptidylpeptidase-IV (Dahyia et al., 1992, Lesuffleur et al., 1990, 1991a); 5-FU-adapted cells are a mixed population of ion-transporting cells and of goblet cells which secrete mucins of colonic immunoreactivity (Lesuffleur et al., 1991c).

(4) The mechanisms of spontaneous and acquired resistance of the cells apparently differ from one drug to another. Amplification of the gene for thymidylate synthase occurs at the first concentration of 5-FU (1x10-6M) the cells are stably adapted to (Lesuffleur et al., 1991c), whereas amplification of dihydrofolate reductase occurs only at the highest concentrations (10-4 and 10-3M) of MTX (Lesuffleur et al., 1991a)

(5) Whatever the drug and concentration the cells are adapted to, they remain, as the parental population, tumorigenic in nude mice.

Altogether these results suggest that, among the parental population, **only the small proportion of cells able to differentiate is able to spontaneously**

escape the cytotoxic effect of MTX and 5-FU. This has been further confirmed by the observation of increased resistance and adaptability to MTX and 5-FU of HT-29 subpopulations selected for their commitment to differentiation (Lesuffleur et al., 1991b).

CONCLUSIONS

Three main conclusions can be drawn from the reported data. A first conclusion, which raises no question, is the ability of colon cancer cell lines to differentiate. A second conclusion is that cells with a similar differentiation potential are also present in tumors. However a considerable effort should be devoted to further analysis of a larger number of tumors. The availability of a large panel of antibodies specific for an increasing number of differentiation-associated markers should allow an easy immunohistological screening of clinical specimens. A precise characterization of differentiated cell types present in tumors should be considered in regard to the phenotype-dependent drug resistance of HT-29 cells. The third conclusion is that, within a heterogeneous population of cancer cells, it is possible to isolate differentiated phenotypes through various conditions of pressure. Indeed the observation that differentiated cell types can be selected under pressure conditions as different as glucose deprivation, sodium butyrate, HMBA, MTX or 5-FU would suggest that drug resistance may be only one aspect of the general resistance properties associated with the commitment of cells to differentiate. Whether common mechanisms are involved in such resistance properties remains to be elucidated. Whether the results obtained with HT-29 cells are specific for this line or a more general phenomenon will also need an extension to other cell lines. It is however interesting to note that the Caco-2 cell line, which is homogeneously and spontaneously differentiated, has been established from a tumor in a patient treated with 5-FU. If our hypothesis is correct it means that these cells are the result of an *in vivo* selection.

ACKNOWLEDGEMENTS

This work was supported partially by the Association pour la Recherche sur le Cancer (ARC), the Fondation pour la Recherche Médicale Française and INSERM (Réseau de Recherche Clinique 489015)

REFERENCES

Audus, K.L., BartelL, R.L., Hidalgo, I.J. & Borchardt, R.T. The use of cultured epithelial and endothelial cells for drug transport and metabolism studies. **Pharm. Res.**, 7:435-451, 1970.

Augeron, C. & Laboisse, C. Emergence of permanently differentiated cell clones in a human colonic cancer cell line in culture after treatment with sodium butyrate. **Cancer Res.**, 44:3961-3969, 1984.

Bear, C.E. & Reyes, E.F. cAMP-activated chloride conductance in the colonic cell line, Caco-2. **Am. J. Physiol.**, 262 (Cell Physiol., 31), C251-C256, 1992.

Blais, A. Expression of Na+-coupled sugar transport in HT-29 cells : modulation by glucose. **Am. J. Physiol.** (Cell Physiol.), 29:C-1245-C1252, 1991.

Blais, A., Bissonnette, P. & Berteloot, A. Common characteristics for Na^+-dependent sugar transport in Caco-2 cells and human fetal colon. **J. Membr. Biol.**, 99:113-125, 1987.

Bouziges, F., Simo, P., Simon-Assmann, P., Haffen, K. & Kedinger, M. Altered deposition of basement-membrane molecules in co-cultures of colonic ancer cells and fibroblast. **Int. J. Cancer**, 48:101-108, 1991.

Brattain, M.G., Brattain, M.D., Fine, W.D., Khaled, F.M., Marks, M.E., Kimball, P.M., Arcolano, L.A. & Danbury, B.H. Initiation and characterization of cultures of human colonic carcinoma with different biological characteristics utilizing feeder layers of confluent fibroblasts. **Oncodev. Biol. Med.**, 2:355-366, 1981.

Carboni, J.M., Howe, C.L., West, A.B., Barwick, K.W., Mooseker, M.S. & Morrow, J.S. Characterization of intestinal brush border cytoskeletal proteins of normal and neoplastic human epithelial cells: a comparaison with the avan brush border. **Am. J. Pathol.,** 129: 589-600, 1987.

Chantret, I., Barbat, A., Dussaulx, E., Brattain, M.G., Zweibaum, A. Epithelial polarity, villin expression and enterocytic differentiation of cultured human colon carcinoma cells : a survey of twenty cell lines. **Cancer Res.**, 48:1936-1942, 1988.

Chantret, I., Lacasa, M., Chevalier, G., Ruf, J., Islam, I., Mantei, N., Edwards, Y., Swallow, D. & Rousset, M. Sequence of the complete cDNA and 5' structure of the human sucrase-isomaltase gene ; homology with a yeast glucoamylase. **Biochem. J.** (1992), in press.

Chantret, I., Trugnan, G., Dussaulx, E., Zweibaum, A. & Rousset, M. Monensin inhibits the expression of sucrase-isomaltase in Caco-2 cells and the mRNA level. **FEBS Lett.**, 235:125-128, 1988.

Crawford, I., Maloney, P.C., Zeitlin, P.L., Guggino, W.B., Hyde, S.C., Turley, H., Gatter, K.C., Harris, A., & Higgins, C.F. Immunocytochemical localization of the cystic fibrosis gene product CFTR. **Proc. Natl. Acad. Sci. USA**, 88:9262-9266, 1991.

Cross H.S. & Quaroni, A. Inhibition of sucrase-isomaltase expression by EGF in the human colon adenocarcinoma cell line Caco-2. **Am. J. Physiol.** 261 (Cell Physiol., 30):C1173-C1183, 1991.

Czernichow, B., Simon-Assmann, P., Kedinger, M., Arnold, C., Parache, M., Marescaux, J., Zweibaum, A. & Haffen, K. Sucrase-isomaltase expression and enterocytic ultrastructure of human colorectal tumors. **Int. J. Cancer**, 44: 238-244, 1989.

Dahiya, R., Lesuffleur, T., Kwak, K.S., Byrd, J.C., Barbat, A., Zweibaum, A. & Kim, Y.S. Expression and characterization of mucins associated with the resistance to methotrexate of HT-29 human colonic adenocarcinoma cell line. **Cancer Res.** (1992), 52:4665-4662.

Dantzig, A.H. & Bergin, L. Uptake of cepholsporin-cephalotoxin, by a dipeptide transport carrier in the human intestinal cell line, Caco-2. **Biochim. Biophys. Acta**, 1027:211-217, 1990.

Darfeuille-Michaud, A., Aubel, D., Chauvière, G., Rich, C., Bourges, M., Servin, A. & Joly, B. Adhesion of enterotoxigenic *Escherichia coli* to the human colon carcinoma cell-line Caco-2 in culture. **Infect. Immun.,** 58:893-902, 1990.

Darmoul, D., Lacasa, M., Baricault, L., Marguet, D., Sapin, C., Troto, P., Barbat, A. & Trugnan, G. Dipeptidylpeptidase IV (CD26) gene expression in enterocyte-like colon cancer cell lines HT-29 and Caco-2. Cloning of the complete human coding sequence and changes of DPP IV mRNA levels during cell differentiation. **J. Biol. Chem.** , 267:4824-4833, 1991.

Daschti, N., Smith, E.A. & Alaujovic, P. Increased production of apolipoprotein B and its lipoproteins by oleic acid in Caco-2 cells. **J. Lipid Res.**, 31:113-123, 1990.

Devedjian, J.C., Fargues, M., Denis-Pouxviel, C., Daviaud, D., Prats, H. & Paris, H. Regulation of the α_{2A}-adrenergic receptor in the HT-29 cell line. Effects of insulin and growth factors. **J. Biol. Chem.**, 266:14359-14366, 1991.

Dix, C.J., Hassan, I.F., Obray, H.Y.,Stah, R. & Wilson, G. The transport of vitamin B12 through polarized monolayers of Caco-2 cells. **Gastroenterology**, 98:1272-1279, 1990.

Dudouet, B., Robine, S., Huet, C., Sahuquillo-Merino, C., Blair, L., Coudrier, E. & Louvard, D. Changes in villin synthesis and subcellular distribution during intestinal differentiation of HT29-18 clones. **J. Cell Biol.**, 105: 359-369, 1987.

Eilers, U., Klumpermann, J. & Hauri, H.P. Nocodazole, a microtubule-active drug, interferes with apical protein delivery in cultured intestinal epithelial cells (Caco-2). **J. Cell Biol.**, 108:13-22, 1989.

Fantini, J., Abadie, B., Tirard, A., Remy, L., Ripert, J.P., El Battari, A. & Marvaldi, J. Spontaneous and induced dome formation by two clonal cell populations derived from a human adenocarcinoma cell line, HT-29. **J. Cell Sci.,** 83: 235-249, 1986.

Fantini, J., Martin, J.M., Luis, J., Rémy, L., Triard A., Marvaldi, J. & Pichon, J. Restricted localization of functional vasoactive intestinal peptide (VIP) receptors in in vitro differentiated human colonic adenocarcinoma cells (HT29-D4). **Eur. J. Cell Biol.**, 46:458-465, 1988.

Faust, R.A. & Albers, J.J. Regulated vectorial secretion of cholesteryl ester transfer protein (LTP-1) by the Caco-2 model of human enterocyte epithelium. **J. Biol. Chem.**, 263:8786-8789, 1988.

Field, F.J., Albright, E. & Mathur, S.N. Regulation of cholesterol esterification by micellar cholesterol in Caco-2 cells. **J. Lipid Res.,** 28:1057-1066, 1987.

Field, F.J., Albright, E. & Matur, S.N. Regulation of triglyceride-rich lipoprotein secretin by fatty acids in Caco-2 cells. **J. Lipid Res.**, 29:1427-1437, 1988.

Fischer, H., Illek, B., Negulescu, P.A., Clauss, W. & Machen, T.E. Carbachol-activated calcium entry into HT-29 cells is regulated by both membrane potential and cell volume. **Proc. Natl. Acad. Sci. USA**, 89:1438-1442, 1992.

Fogh, J. & Trempe, G. New human tumor cell lines. In : J. Fogh (ed.), **Human Tumor Cells "In vitro"**, pp. 115-141, New York : Plenum Publ. Corp. 1975.

Fogh, J., Fogh J.M. & Orfeo, T. One-hundred and twenty-seven cultured human tumor cell lines producing tumors in nude mice. **J. Natl. Cancer Inst.**, 59:221-225, 1977.

Gaillard, J.L., Berche, P., Frehel, C., Govin E., & Govin, P. Entry of L-monocytogenes into cells is mediated by internalin, a repeat protein reminiscent of surface antigens from Gram-positive cocci. **Cell**, 65:1127-1141, 1991.

Gilbert, T., Le Bivic, A., Quaroni, A. & Rodriguez-Boulan, E. Microtubular organization and its involvement in the biogenetic pathways of plasma membrane proteins in Caco-2 intestinal epithelial cells. **J. Cell Biol.**, 113:275-288, 1991.

Giuliano, A.R. & Wood, R. Vitamin D-regulated calcium transport in Caco-2 cells : unique *in vitro* model. **Am. J. Physiol.**, 261 (Gastrointest. Liver Physiol., 23), G207-G212, 1991.

Giuliano, A.R., Franceschi R.T. & Wood, R.J. Characterization of the vitamin D receptor from the Caco-2 human colon carcinoma cell line : effect of cellular differentiation. **Arch. Biochem. Biophys.**, 285:261-269, 1991.

Godefroy, O., Huet, C., Ibarra, C., Dautry-Varsat, A. & Louvard, D. Establishment of polarized endocytosis in differentiable intestinal HT29-18 subclones. **The New Biologist**, 2: 875-886, 1990.

Grasset, E., Pinto, M., Dussaulx, E., Zweibaum, A. & Desjeux, J.F. Epithelial properties of human colonic carcinoma cell line Caco-2 : electrical parameters. **Am. J. Physiol.** 247 (Cell. Physiol., 16): 260-267, 1984.

Hafez, M.M., Infante, D., Winawer, S. & Friedman, E. Transforming growth factor β_1 acts as an autocrine-negative growth regulator in colon enterocytic differentiation but not in goblet cell maturation. **Cell Growth & Diff.**, 1:617-626, 1990.

Hauri, H.P., Sterchi, E.E., Bienz, D., Fransen, J.A.M. & Marxer, A. Expression and intracellular transport of microvillus membrane hydrolases in human intestinal epithelial cells. **J. Cell Biol.**, 101:838-851, 1985.

Hawck, W. & Stanners, C.P. Control carcinoembryonic antigen gene family expression in a differentiating colon carcinoma cell line, Caco-2. **Cancer Res.**, 51:3526-3533, 1991.

Hidalgo, I.J. & Borchardt, R.T. Transport of bile acids in a human intestinal epithelial cell line, Caco-2. **Biochem. Biophys. Acta,** 1035:97-103, 1990.

Hilgers, A.R. & Barton, P. Human colon carcinoma cell monolayers as a model for drug transport across the intestinal epithelium. I. Morphological and permeability characteristics. **Pharm. Res.**, 5:PD947, 1988.

Huet, C., Sahuquillo-Mérino, C., Coudrier, E. & Louvard, D. Absorptive and mucus-secreting subclones isolated from a multipotentent intestinal cell line (HT-29) provide new models for cell polarity and terminal differentiation. **J. Cell Biol.**, 105:345-358, 1987.

Hughes, T.E., Ordovas, J.M. & Schaeffer, E.J. Regulation of intestinal apolipoprotein B synthesis and secretion by Caco-2 cells. Lack of fatty acid affects and control by intracellular calcium ion. **J. Biol. Chem.**, 263:3425-3431, 1988.

Hughson, E.J. & Hopkins, C.R. Enterocytic pathways in polarized Caco-2 cells: identification of an endosomal compartment accessible from both apical and basolateral surfaces. **J. Cell Biol.**, 110:337-348, 1990.

Kaiserlian, D., Rigal, D., Abello, J. & Revillard, J.P. Expression, function and regulation of the intercellular adhesion molecule -1 (KAM-1) on humanintestinal epithelial cell lines. **Eur. J. Immunol.**, 21:2415-2421, 1991.

Kreusel, K.M., Fromm, M., Schulzke, J.D. & Hegel, V. Cl- secretion in epithelial monolayers of mucus-forming human colon cells (HT-29/B6). **Am. J. Physiol.** 247 (Cell. Physiol., 30): C574-C582, 1991.

Laboisse, C. Maoret, J.J., Triadou, N. & Augeron, C. Restoration by polyethylene glycol of characteristics of intestinal differentiation in subpopulations of the human colonic adenocarcinoma cell line HT-29. **Cancer Res.**, 48:2498-2594, 1988.

Laburthe, M., Rousset, M., Rouyer-Fessard, C., Couvineau, A., Chantret, I., Chevalier, G. & Zweibaum, A. Development of vasoactive intestinal peptide-responsive adenylate cyclase during enterocytic differentiation of Caco-2 cells in culture. Evidence for an increased receptor level. **J. Biol. Chem.**, 262:10180-10184, 1987.

Lacroix, B., Kedinger, M., Simon-Assmann, P., Rousset, M., Zweibaum, A. & Haffen, K. Develomental pattern of brush border enzymes in the human fetal colon. Correlation with some morphogenetic events. **Early Human Develop.**, 9:95-103, 1984.

Le Bivic, A., Quaroni, A., Nichols, B. & Rodriguez-Boulan, E. Biogenetic pathways of plasma membrane proteins in Caco-2, a human intestinal epithelial cell line. **J. Cell Biol.**, 111:1351-1361, 1990.

Lesuffleur, T., Barbat, A., Dussaulx, E. & Zweibaum, A. Growth adaptation to methotrexate of HT-29 human colon carcinoma cells is associated with their ability to differentiate into columnar absorptive and mucus-secreting cells. **Cancer Res.**, 50: 6334-6343, 1990.

Lesuffleur, T., Barbat, A., Luccioni, C., Beaumatin, J., Claire, M., Kornowski, A., Dussaulx, E., Dutrillaux, B. & Zweibaum, A. Dihydropholate reductase gene amplification-associated shift of differentiation in methotrexate-adapted HT-29 cells. **J. Cell. Biol.**, 115: 1409-1418,1991a.

Lesuffleur, T., Kornowski, A., Augeron, C., Dussaulx, E., Barbat, A., Laboisse, C. & Zweibaum, A. Increased growth adaptability to 5-fluorouracil and methotrexate of HT-29 subpopulations selected for their commitment to differentiation. **Int. J. Cancer,** 49:731-737, 1991b.

Lesuffleur, T., Kornowski, A., Luccioni, C., Muleris, M., Barbat, A., Beaumatin, J., Dussaulx, E., Dutrillaux, B. & Zweibaum, A. Adaptation to 5-fluorouracil of the heterogenous human colon tumor cell line HT-29 results in the selection of cells committed to differentiation. **Int. J. Cancer,** 49:721-730,1991c.

Mahraoui, L., Rousset, M., Dussaulx, E., Darmoul, D., Zweibaum, A. & Brot-Laroche, E. Expression of the facilitative hexose transporter, GLUT5, in the human colon cancer cell line Caco-2 and human small intestine and colon. **Am. J. Physiol.,** 263 (Gastrointest. Liver Physiol. 26):G312-G318, 1992.

Matsumoto, H., Erickson, R.J., Gum, J.R., Yoshioka, M., Gum, E. & Kim, Y.S. Biosynthesis of alkaline phosphatase during differentiation of the human colon cancer cell line Caco-2. **Gastroenterology**, 98:1199-1207, 1990.

Matter, K., Brauchbar, M., Bucher, K. & Hauri, H.P. Sorting of endogenous plasma membrane proteins occurs from two sides in cultured human intestinal epithelial cells (Caco-2). **Cell**, 60:429-437, 1990a.

Moberly, J.B., Cole, T.G., Alpers, D.H. & Schonfeld, G. Oleci acid stimulation of apolipoprotein B secretion from Hep G2 and Caco-2 cells occurs post-transcriptionally. **Biochim. Biophys. Acta.**, 1042:70-80, 1990.

Moll, R., Robine, S., Dudouet, B. & Louvard, D. Villin: a cytoskeletal protein and a differentiation marker expressed in some human adenocarcinomas. **Virchows Arch.**, 54: 155-169, 1987.

Montrose-Rafizadeh, C., Guggino, W.B., & Montrose, M.H. Cellular differentiation regulates expression of Cl- transport and cystic fibrosis transmembrane conductance regulator mRNA in human intestinal cells. **J. Biol. Chem.**, 266:4495-4499, 1991.

Phillips, T.E., Huet, C., Bilbo, P.R., Podolsky, D.K., Louvard, D., & Neutra, M.R. Human intestinal goblet cells in monolayer culture : characterization of a mucus-secreting subclone derived from the HT-29 colon adenocarcinoma cell line. **Gastroenterology**, 94:1390-1442, 1988.

Pinto, M., Appay, M.D., Simon-Assmann, P., Chevalier, G., Dracopoli, N., Fogh, J. & Zweibaum, A. Enterocytic differentiation of cultured human colon cancer cells by replacement of glucose by galactose in the medium. **Biol. Cell.,** 44:193-196, 1982.

Pinto, M., Robine-Léon, S., Appay, M.D., Kedinger, M., Haffen, K., Fogh, J. & Zweibaum, A. Enterocyte-like differentiation and polarization of the human colon carcinoma cell line Caco-2 in culture. **Biol. Cell.,** 47:323-330, 1983.

Pringault, E., Arpin, M., Garcia, A., Finidoril, J. & Louvard, D. A human villin cDNA clone to investigate the differentiation of intestinal and kidney cells in vivo and in culture. **EMBO J.**, 5:3119-3124, 1986.

Ramanujam, K.S., Seetharam, S., Ramasamy, M. & Seetharam, B. Expression of cobalamin transport proteins and cobalamin transcytosis by colon adenocarcinoma cells. **Am. J. Physiol.,** 260 (Gastrointest. Liver Physiol, 23):G416-G422, 1991.

Real, F.X., Xu, M., Vilà, M.R., de Bolos, C. Intestinal brush border associated enzymes : coordinate expression in colorectal cancer. **Int. J. Cancer,** 1992 (in press).

Rindler, M.J. & Traber, M.G. A specific sorting signal is not required for the polarized secretion of newly synthesized proteins from cultured intestinal epithelial cells. **J. Cell Biol.**, 107:471-479, 1988.

Robine, S., Huet, C., Moll, R., Sahuquillo-Merino, C., Coudrier, E., Zweibaum, A. & Louvard, D. Can villin be used to identify malignant and undifferentiated normal digestive epithelial cells? **Proc. Natl. Acad. Sci. USA**, 82: 8488-8492, 1985

Rousset, M., Chantret, I., Darmoul, D., Trugnan, G., Sapin, C., Green, F., Swallow, D. & Zweibaum, A. Reversible forskolin-induced impairment of sucrase-isomaltase mRNA levels, biosynthesis and transport to the brush border membrane in Caco-2 cells. **J. Cell Physiol.**, 141:627-635, 1989.

Sood, R., Bear, C., Auerbach, W., Reyes, E., Jensen, T., Kartner, N., Riordan, J.R. & Buchwald, M. Regulation of CFTR expression and function during differentiation of intestinal epithelial cells. **EMBO J.**, 11:2487-2494, 1992.

Tilly, B.C., Kansen, M., Van Gageldonk, G.M., Van Den Berghe, N., Galjaard, H., Bijman, J. & De Jonge, H.R. G-proteins mediate intestinal chloride channel activation. **J. Biol. Chem.**, 266: 2036-2040, 1991.

Trotter, P.J. & Storch, J. Fatty acid uptake and metabolism in a human intestinal cell line (Caco-2) : comparison of apical and basolateral incubation. **J. Lip. Res.**, 32:293-304, 1991.

Van't Hoff, W. & Van Meer, G. Generation of lipid polarity in intestinal epithelial (Caco-2) cells : sphingolipid synthesis in the Golgi complex and sorting before vesicular traffic to the plasma membrane. **J. Cell Biol.**, 111:977-986, 1990.

Viallard, V., Lacombe, B., Trocheris, V., Tabacik, C., Aliau, S. Metabolism of low-density lipoprotein in differentiated and undifferentiated HT-29 colon cancer cells. **Int. J. Cancer**, 46:320-325, 1990.

Watson, A.J.M., Levine, S., Donowitz, M. & Montrose, M.H. Kinetics and regulation of a polarized Na^+-H^+ exchanger from Caco-2 cells, a human intestinal cell line. **Am. J. Physiol.**, 261 (Gastrointest. Liver Physiol., 24), G229-G238, 1991.

Wice, B.M., Trugnan, G., Pinto, M., Rousset, M., Chevalier, G., Dussaulx, E., Lacroix, B. & Zweibaum, A. The intracellular accumulation of UDP-N-acetylhexosamines is concomitant with the inability of human colon cancer cells to differentiate. **J. Biol. Chem.**, 260:139-146, 1985.

Wilson, G., Hassan, I.F., Dix, C.J., Williamson, I., Stah, H., MacKay, M. & Artursson, P. Transport and permeability properties of human Caco-2 cells : an *in vitro* model of the intestinal epithelial cell barrier. **J. Controlled Release**, 11, 25-40, 1990.

Wiltz, O., O'Hara, C.J., Steele, G.D. & Mercurio, A.M. Sucrase-isomaltase: a marker associated with the progression of adenomatous polyps to adenocarcinomas. **Surgery**, 108: 269-276, 1990.

Woodcock, S., Williamson, I., Hassan, I. & MacKay, M. Isolation and characterization of clones from the Caco-2 cell line displaying increased taurocholic acid transport. **J. Cell Sci.**, 98:323-332, 1991.

Zeitlin, P.L., Crawford, I., Lu, L., Woel, S., Cohen, M.E., Donowitz, M., Montrose, M.H., Hamosh, A., Cutting, G.R., Gruenert, D., Huganir, R., Maloney, P. & Guggino, W.B. CFTR protein expression in primary and cultured epithelia. **Proc. Natl. Acad. Sci. USA**, 89:344-347, 1992.

Zweibaum, A., Hauri, H.P., Sterchi, E., Chantret, I., Haffen, K., Bamat, J. & Sordat, B.. Immunohistological evidence obtained with monoclonal antibodies of small intestinal brush border hydrolases in human colon cancers and foetal colons. **Int. J. Cancer,** 34:591-598, 1984.

Zweibaum, A., Pinto, M., Chevalier, G., Dussaulx, E., Triadou, N., Lacroix, B., Haffen, K., Brun, J.L. & Rousset, M. Enterocytic differentiation of a subpopulation of the human colon tumor cell line HT-29 selected for growth in sugar-free medium and its inhibition by glucose. **J. Cell. Physiol.**, 122:21-29, 1985.

Zweibaum, A., Triadou, N., Kedinger, M., Augeron, C., Robine-Léon, S., Pinto, M., Rousset, M. & Haffen, K. Sucrase-isomaltase : a marker of foetal and malignant epithelial cells of the human colon. **Int. J. Cancer**, 32:407-412, 1983.

THE CONTROL OF CELL MULTIPLICATION AND DIFFERENTIATION IN HUMAN MYELOMONOCYTIC CELLS

E. Huberman, D.A. Tonetti, M. Horio, S. Murao, and F.R. Collart
Biological and Medical Research Division
Argonne National Laboratory
9700 South Cass Avenue
Argonne, IL 60439

Abstract

To identify steps in the signal transduction pathway that lead to monocyte/macrophage maturation, we used human promyelocytic HL-60 cells that are either susceptible or resistant to such a differentiation induced by phorbol 12-myristate 13-acetate (PMA). Unlike the cells from susceptible cell lines, the PMA-resistant variants did not express the genes that code for the protein kinase C (PKC) β isozyme and a "δ-like" PKC. The resistant cells also exhibited an attenuated PMA-induced expression of the early response genes *junB*, c-*fos*, and c-*jun*. These results suggest that the signal transduction pathway that leads to PMA-induced differentiation in the HL-60 cell system requires specific PKC isozymes for the proper expression of the early response genes and ultimately for the expression of genes that define the mature state.

We have also examined the role a protein complex (PC) composed of a 10- and a 14-kDa protein plays in regulating cell multiplication during terminal differentiation of myelomonocytic cells. The genes coding for these proteins are expressed during the late stages of chemically induced terminal differentiation in two human myelomonocytic leukemia cell lines, HL-60 and THP-1. This expression is associated with terminal differentiation because incubation of HL-60 cells with an agent or condition that causes suppression of cell growth but not induction of differentiation does not result in expression of the PC. At concentration of 5–15 nM, the purified PC inhibited the growth of HL-60 and THP-1 cells as well as that of other cell types belonging to different cell lineages. This growth inhibition was preceded by a reduction in [^{32}P]phosphate incorporation. The specific expression pattern and growth-inhibitory character of the PC suggest that the complex may have a role in suppressing cell growth during myelomonocytic terminal differentiation induced by specific chemical stimuli and during physiological and pathological events associated with myelomonocytic cell functions.

NATO ASI Series, Vol. H 75
Cancer Therapy
Edited by N. D'Alessandro, E. Mihich, L. Rausa, H. Tapiero, and T. R. Tritton

Introduction

Human cells produce and respond to growth- and differentiation-inducing factors. Stem cells begin to mature after the interaction of a specific inducer of differentiation with its appropriate cellular receptor. Shortly following this interaction, a series of cellular signals is transmitted from the receptor to the genome, causing the activation and expression of a series of genes, regulatory genes in particular. The products of these early activated genes cause, through positive or negative controls (e.g., through a "trans"-acting process or ligand-receptor interaction), sequential expression of genes that code for the different functions associated with the mature state (including inhibition of cell multiplication). This paper will first deal with the role of a specific protein kinase in the signal transduction processes that lead to differentiation induction in human promyelocytic leukemia cells and then will describe the role a specific protein complex may play in regulating multiplication during human myelomonocytic cell differentiation.

Part I. Signal Transduction Processes Leading to Phorbol Ester-Induced Differentiation in Human Leukemia Cells

Protein kinase C (PKC), the receptor for the tumor promoter phorbol 12-myristate 13-acetate (PMA), is a serine/threonine-specific protein kinase which plays a central role in cellular signal transduction processes (Nishizuka 1988). PKC is comprised of a family of eight isozymes (α, β_I, β_{II}, γ, δ, ε, ζ, and η/L) that have been cloned and characterized biochemically (Blumberg 1991). The β_I and β_{II} isozymes are derived by alternative splicing of the same gene and differ in 50 amino acids at their carboxyl ends. The PKCs can be subdivided into two groups based on structure and function. Group I (α, β_I, β_{II}, and γ) and Group II (δ, ε, ζ, and η/L) isozymes exhibit extensive sequence similarity within each group but display less similarity between the groups (Huang 1990). The structural diversity observed between the two groups is reflected in substrate specificity and cofactor requirements; Group I isozymes are Ca^{2+}-dependent and the Group II isozymes are Ca^{2+}-independent.

Human promyelocytic HL-60 leukemia cells are susceptible to induction of a monocytic/macrophage differentiation by a variety of chemicals including PMA. To decipher some of the signal transduction events that are involved in such a differentiation, we isolated HL-60 leukemia cell variants that differ in their response to PMA (Homma et al. 1986). Variants HL-525 and HL-534 are resistant to the induction of differentiation by PMA, yet remain susceptible to other inducers of monocytic and granulocytic differentiation such as 1α,25-dihydroxyvitamin D_3 [1α,25$(OH)_2D_3$] and retinoic acid. Cells from another variant, HL-205, which is clonally derived from HL-60, are as susceptible as the parental cells to differentiation induction by PMA and the other inducers. Unlike the PMA-susceptible cells, the PMA-resistant HL-525 and HL-534 cells are incapable of translocating PKC from the cytosol to the membrane fraction and demonstrate differential phosphorylation of endogenous proteins following PMA treatment (Homma et al. 1986). Sensitive and resistant cell extracts exhibit PKC activities that differ in their requirements for Ca^{2+} and phospholipid cofactors and in their specificity toward certain defined peptide substrates (Homma et al. 1988). These observations suggest that the PMA-resistant cells differ from the PMA-susceptible cells in their PKC isozyme profile.

Expression of PKC Isoenzyme Genes in HL-60 Cells Susceptible and Resistant to PMA-Induced Differentiation. To investigate the possibility that PMA resistance is due to a difference in PKC profile, we examined the PKC isozyme steady-state mRNA levels in the PMA-susceptible and -resistant cells by Northern blot analysis. Total RNA or poly(A)$^+$ RNA was isolated from HL-60, HL-205, HL-525, and HL-534 cells and hybridized with the human PKC α, β, and γ cDNA or with a specific oligonucleotide probe. Equal loading of RNA was assessed by hybridization with a glyceraldehyde phosphate dehydrogenase (GAPD) cDNA probe. The PKC β cDNA probe and a PKC β_{II}-specific oligonucleotide probe each recognized transcripts of 3.7 and 8.7 kb. Whereas the level of the 3.7-kb transcript was 4- to 5-fold more abundant in the PMA-susceptible HL-60 and HL-205 cells than in the PMA-resistant HL-525 and HL-534 cells, the level of the 8.7-kb transcript was 7- to 9-fold more abundant (Figs. 1*A* and 1*B*). We observed the same PKC β transcript levels using either total or poly(A)$^+$ RNA. Hybridization with two β_I-specific oligonucleotide probes detected a barely visible 8.7-kb transcript in the HL-60 and HL-205 cells, and almost none in the HL-525 and HL-534 cells. After hybridization of a PKC α cDNA probe with poly(A)$^+$ RNA, similar levels of a 3.2-kb and a 3.7-kb transcript were observed in the four cell types, while a larger transcript of 9.5 kb was observed mainly in HL-205 cells (Fig. 1*C*). Although the 3.2-kb transcript was detected in total RNA, it was more distinctly visualized using poly(A)$^+$ RNA. Comparable results were obtained using a PKC α oligonucleotide probe. The reduced levels of PKC α transcripts in the HL-534 cells are attributed to a lower amount of RNA loaded in the lane (as the GAPD level indicates; Fig. 1*C*) and are in agreement with replicate experiments which demonstrated similar levels of PKC α in all four cell lines. None of the four cell types exhibited detectable levels of PKC γ transcripts.

Human cDNAs corresponding to the rat Ca^{2+}-independent PKC δ, ε, and ζ isozymes have not yet been cloned. We therefore used oligonucleotide probes specific to these rat PKCs to investigate expression of the PKC δ, ε, and ζ genes in HL-60, HL-205, HL-525,

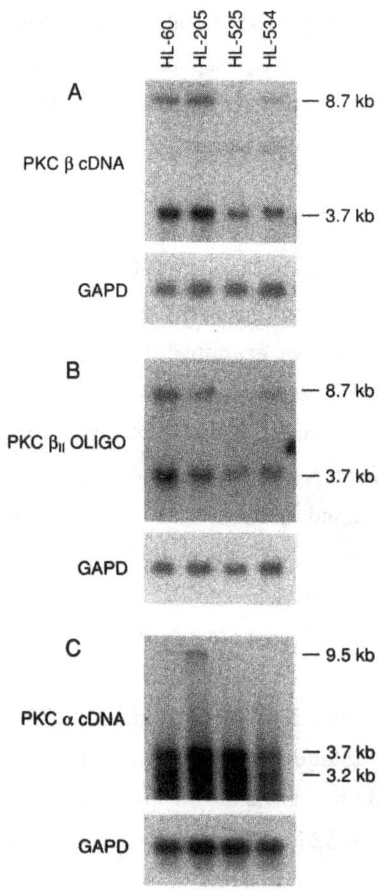

Fig. 1. Northern blot analysis of PKC isozyme RNA transcripts in PMA-sensitive and resistant cells. A, total RNA hybridized to the human PKC β cDNA probe; B, total RNA hybridized to a human PKC β_{II} oligonucleotide probe; C, poly(A)$^+$ RNA hybridized to the human PKC α cDNA probe. Hybridization to the GAPD cDNA probe is shown beneath each lane to assess the quantity of RNA loaded per lane. (From Tonetti et al. 1992)

and HL-534 cells. The rat PKC δ oligonucleotide probe detected a 3.1-kb transcript in all four cell lines. Although different levels of a "δ-like" transcript were detected in all four cell lines, this transcript was more abundant in the PMA-susceptible cells than in the PMA-resistant HL-525 cells. The rat PKC ε and ζ oligonucleotide probes did not detect specific transcripts in the four cell types. These results indicate that resistance to PMA-induced differentiation in HL-525 and HL-534 cells is associated with a reduction in the levels of the PKC β_{II} and "δ-like" PKC transcripts.

To determine if the differences in specific PKC RNA levels were reflected in the corresponding cellular enzyme levels, we examined the relative amounts of the individual PKC isozymes by immunoblot analysis. No substantial difference in the level of PKC α was detected in the four cell types (Fig. 2). The PMA-resistant HL-525 and HL-534 cell lines contained at least an order of magnitude less PKC β protein than the PMA-susceptible HL-60 and HL-205 cell lines (Fig. 2*B*). Since the PKC β monoclonal antibody does not discriminate between β_I and β_{II} isozymes, the observed deficit in PKC β protein in the resistant cells may reflect a reduction in either β_I or β_{II} isozymes. The PKC γ isozyme protein was not detected in the four cell types. Additionally, we were unable to determine the relative levels of PKC δ protein because the available polyclonal antibody against the rat PKC δ enzyme gave inconclusive results. These studies indicate that the

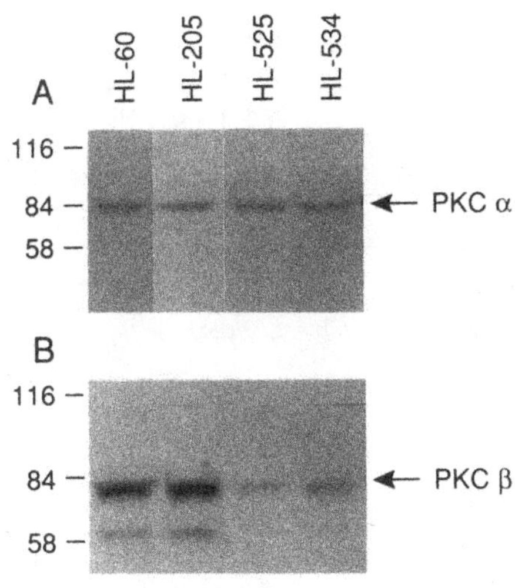

Fig. 2. Immunoblot analysis of PKC α and β in PMA-susceptible HL-60 and HL-205 cells and in PMA-resistant HL-525 and HL-534 cells. A, immunoblot with a PKC α monoclonal antibody; B, immunoblot with a PKC β monoclonal antibody. Positions of molecular weight standards are indicated in kDa on the left. (From Tonetti et al. 1992)

expression of the PKC β gene in the resistant cells is reduced at both the steady-state mRNA and protein levels.

PMA-Induced Expression of Early Response Genes in the PMA-Sensitive and -Resistant Cells. Changes in the expression of the early response genes are associated with PMA-induced differentiation in HL-60 cells (Szabo et al. 1991). To substantiate this association, we analyzed the expression of three early response genes, *junB*, c-*fos* and c-*jun*, in HL-60 and HL-205 cells and in HL-525 and HL-534 cells following treatment with 30 nM PMA. An 18S ribosomal RNA probe was used to standardize the amount of RNA loaded per lane, because the expression of the GAPD gene was found to be modulated by PMA treatment. In the untreated cells, the steady-state level of *junB* RNA was 2- to 3-fold greater in the resistant cells than in the susceptible cells. After 1 hour of treatment with PMA, expression of the *junB* gene was increased in all four cell types, with maximal expression 2–4 hours after treatment (Fig. 3). However, there was a marked difference in the magnitude of the response between the PMA-susceptible and PMA-resistant cell types. In the susceptible HL-60 and HL-205 cells, treatment with PMA at 4 and 9 hours caused a 40- to 50-fold increase in the steady-state level of *junB* RNA, compared with about a 5-fold increase in the HL-525 and HL-534 cells.

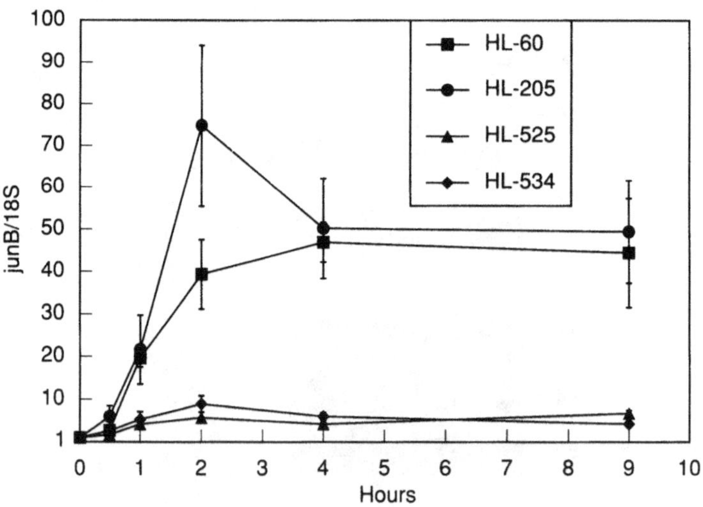

Fig. 3. Abundance of the *junB* RNA transcript, defined by the ratio of *junB* to 18S RNA levels, at different times after treatment of HL-60, HL-205, HL-525, and HL-534 cells with 30 nM PMA. (From Tonetti et al. 1992)

Baseline steady-state levels of c-*fos* RNA were faintly detected in all four cell types. Treatment with 30 nM PMA resulted in an increase in the 2.2-kb c-*fos* transcript. The expression of the c-*fos* gene in all four cell types was biphasic, with maximal expression at 1–2 and 9 hours. The maximum level of c-*fos* gene expression in the PMA-susceptible cells was markedly higher than in the PMA-resistant cells: at 1 hour after PMA treatment, the level of the c-*fos* transcript was 4- to 5-fold higher in HL-60 and HL-205 cells.

PMA also caused an induction of the c-*jun* gene expression in the susceptible cells that was highest at 9 hours after PMA treatment, and the level of induction was higher in HL-205 than in HL-60 cells. Unlike in the PMA-sensitive cells, no detectable increase in the level of the c-*jun* transcript was observed after PMA treatment of HL-525 and HL-534 cells, although the baseline level of this transcript was higher in HL-534 cells than in other cell types. These results indicate that PMA induced the expression of the early response genes *junB*, c-*fos,* and c-*jun* in PMA-susceptible HL-60 and HL-205 cells, and this induction was reduced or absent in HL-525 and HL-534 cells.

Role of Protein Kinase C β in Signal Transduction Processes Leading to PMA-Induced Differentiation. Normally, in response to PMA treatment of HL-60 cells, there is a rapid induction of the expression of c-*jun*, c-*fos* and *junB* genes (Szabo et al. 1991). Our studies indicate an attenuation of this induction and higher baseline levels of the *junB*, c-*fos*, and c-*jun* transcripts in the PMA-resistant cell lines. Since we demonstrated that these variants are deficient in the expression of the PKC β and possibly a "δ-like" PKC gene, these PKC isozymes may be directly involved in the proper expression of the *junB*, c-*fos*, and c-*jun* genes. Alternatively, since the baseline levels of these early response genes are elevated in the PMA-resistant cell lines, the diminished induction may reflect a saturation of the induction process, possibly by a PKC-independent signaling pathway. However, in the case of *junB*, the 2- to 3-fold higher baseline level in the PMA-resistant cells is still not sufficient to account for the level of induction observed in the PMA-sensitive cells. Abnormalities in the levels of Jun and Fos-related proteins were reported in PMA-resistant mouse thymoma cells deficient in PKC ε (Homan et al. 1991, Jensen et al. 1991). Certainly, a defect in the expression of AP-1 proteins will disrupt the delicate balance of factors which determines the formation of Fos/Jun heterodimers, Jun homodimers, and the abundance of the negative regulator JunB. These possibilities can affect the binding affinity for the PMA responsive element and interfere with the transcription of the differentiation-dependent genes (Halazonetis et al. 1988; Chiu et al. 1989; Schütte et al. 1989).

Based on previous and current studies with HL-60 cells that are susceptible or resistant to PMA-induced differentiation, we suggest a signal transduction pathway that results in terminal differentiation of these and related cells. Central to this pathway is the interaction of PMA and related ligands with specific cellular receptors, the PKC isozymes. In the case of the HL-60 cells, PMA interaction with PKC β or PKC "δ-like" isozymes causes their activation and translocation from the cytosol to the membrane fraction (Homma et al. 1988), which also contains nuclear membrane components. The activated PKC isozymes subsequently phosphorylate specific proteins, including nuclear proteins (Anderson et al. 1985), and the phosphorylated nuclear proteins in turn cause the expression of the early response genes. The products of these genes then elicit a cascade of genes, some of which code for cellular functions that induce the mature phenotype.

Part 2. A Protein Complex Expressed During Terminal Differentiation of Myelomonocytic Cells is an Inhibitor of Growth

To identify myelomonocytic cell differentiation markers, we developed a set of murine monoclonal antibodies (mAbs) directed against nuclear antigens (Murao et al. 1985). One of these antibodies, NM-6, was used to analyze the distribution of its antigen in mature and immature blood cells (Murao et al. 1989). The NM-6 antigen was detected in peripheral blood monocytes and granulocytes but not in lymphocytes, and only a limited expression was detected in human myeloid leukemia cell lines (including the HL-60 cells). These results indicated that the NM-6 antigen is a maturation marker in myelomonocytic cells.

To characterize the NM-6 antigen, we purified it from a human spleen with the aid of an NM-6 immunoaffinity column. Denaturing gel electrophoresis indicated that the antigen is a protein complex (PC) composed of equal amounts of two proteins having molecular masses of 10 and 14 kDa (Murao et al. 1989). Using a polyclonal antibody directed against the purified 14-kDa protein, we isolated and sequenced a cDNA clone from a λgt11 human monocyte cDNA library (Murao et al. 1989). The DNA sequence of this clone was the same as that of a cDNA clone that codes for the protein MRP-14 (Dorin et al. 1987; Odink et al. 1987) except for three nucleotides in the 3′ noncoding region. The amino terminus of the 14-kDa protein was blocked, but analysis of an internal peptide derived by V8 protease digestion yielded a 25-amino acid segment that corresponded to a coding segment in the human cDNA clone. Amino-terminal sequence analysis indicated that the 10-kDa protein was likely identical to another protein termed MRP-8 (Odink et al. 1987). Although the NM-6 mAb recognizes the intact PC, it does not recognize purified MRP-8 or MRP-14 constituent proteins.

Expression of the Protein Complex during Differentiation of Human Promyelocytic HL-60 and Monocytic THP-1 Leukemia Cells. To investigate the regulation of the PC during myelomonocytic cell growth and maturation, we examined the expression of its constituent MRP-8 and MRP-14 genes during chemically induced terminal differentiation of HL-60 and THP-1 cells (Murao et al. 1985). The expression of the PC antigen in HL-60 and THP-1 cells was measured by its reactivity with the NM-6 mAb. Less than 3% of untreated HL-60 or THP-1 cells cultured for up to 5 days reacted with the NM-6 mAb. However, the treatment of HL-60 cells with $1\alpha,25(OH)_2D_3$ resulted in a time- and dose-dependent increase in the percentage of reactive cells. The response of the PMA-treated THP-1 cells was more complex: the percentage of positive cells increased with PMA concentrations up to 3 nM and then the percentage declined. In both cell types, treatment with the differentiation inducers resulted in a time- and dose-dependent reduction in cell growth (Fig. 4).

The pattern of PC antigen expression in untreated or treated HL-60 and THP-1 cells was similar to that observed for the steady-state mRNA levels of the PC constituents. Untreated human HL-60 or THP-1 cells expressed little or no MRP-8 (0.55 kbp) and MRP-14 (0.75 kbp) transcripts. The treatment of HL-60 cells with $1\alpha,25(OH)_2D_3$ or THP-1 cells with PMA, which induce a mature monocytic phenotype, caused a marked and coordinated increase in the steady-state levels of both the MRP-8 and MRP-14 mRNAs. Maximal levels of the transcripts were observed between 2 and 4 days after treatment.

To verify that the presence of the PC is associated with terminal differentiation and not merely growth inhibition, we analyzed the expression of the PC antigen in HL-60 cells treated with inducers of differentiation such as $1\alpha,25(OH)_2D_3$ or retinoic acid (RA), with the growth inhibitor VM-26 (teniposide, a specific topoisomerase II inhibitor), and following incubation of the cells in growth medium containing reduced levels of serum. Each of the treatments resulted in a dose-dependent decrease in cell number by the fourth day of treatment. Cells treated with $1\alpha,25(OH)_2D_3$ or RA displayed a mature phenotype as revealed by an increased reactivity of cells with the OKM1 mAb, which

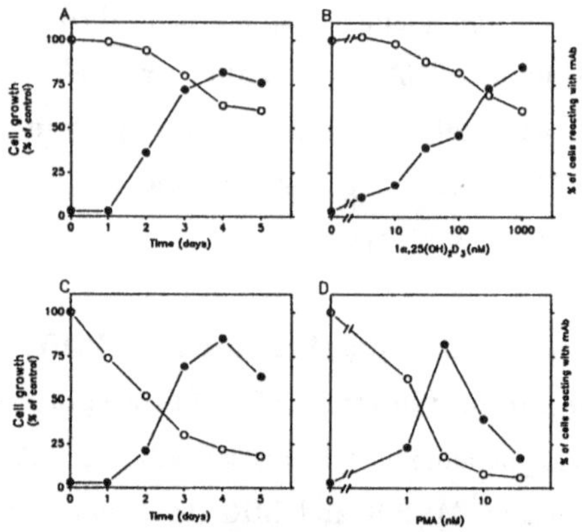

Fig. 4. Time course and dose response profiles for HL-60 (A,B) and THP-1 (C,D) cells treated with differentiation inducers. The number of cells as a percentage of controls (○) and the percentage of cells reacting with the NM-6 mAb (●) were determined over time (A,C) after treatment with 400 nM $1\alpha,25(OH)_2D_3$ (HL-60) or 3 nM PMA (THP-1). For the dose response (B,D), the same variables were determined after 4 days of treatment. (From Murao et al. 1990)

detects a surface antigen common to mature monocytes and granulocytes (Foon et al. 1982). Treatment with VM-26 or reducing the serum concentrations in the growth medium caused growth inhibition but did not increase cell maturation (as determined by increased reactivity with the OKM1 mAb), nor did it yield an increase in reactivity with the NM-6 mAb. Reactivity with the NM-6 mAb was limited to cultures that reacted with the OKM1 mAb and exhibited a decrease in the number of cells. These results indicate that the PC is expressed during terminal differentiation but not during growth inhibition that is not associated with such differentiation.

Secretion of the PC from Differentiating HL-60 Cells. To determine whether the PC was secreted from the differentiating HL-60 cells, we determined the amount of the PC in untreated cells, in cells treated with $1\alpha,25(OH)_2D_3$, and in the culture media from these cells. [^{35}S]Methionine-labeled MRP-8 and MRP-14 proteins were detected after immunoprecipitation of the PC from extracts of $1\alpha,25(OH)_2D_3$-treated HL-60 cells (Fig. 5, lane 3) and from such cells following treatment with 5 nM PMA for 4 h prior to analysis (to facilitate secretion of the PC; Fig. 5, lane 4). The relative band intensities corresponded to the relative amount of methionine in each protein (MRP-14 has 3 times as many methionine residues as MRP-8). Labeled MRP-8 and MRP-14 proteins were also detected in culture media from cells treated with both $1\alpha,25(OH)_2D_3$ and PMA (Fig. 5, lane 8) but not in extracts of untreated HL-60 cells,

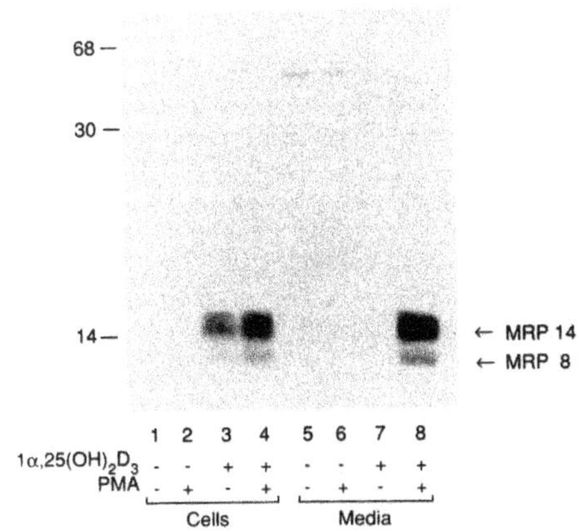

Fig. 5. Analysis of [^{35}S]methionine-labeled protein complex in 1α,25(OH)$_2$D$_3$-treated HL-60 cells and culture media. HL-60 cells were treated with 1 μM 1α,25(OH)$_2$D$_3$, and the amount of the PC was determined by immunoprecipitation of the label proteins. The immunoprecipitated proteins from cells (lanes 1–4) or culture media (lanes 5–8) were electrophoresed through a sodium dodecyl sulfate-polyacrylamide gel and the labeled MRP-8 and MRP-14 proteins were detected by autoradiography of the dried gel. The treatment is indicated at the bottom. (From Murao et al. 1990)

in cells treated for 4 h with only PMA, or in their culture media (Fig. 5, lanes 1, 2, 5, and 6). These results indicate that the PC is secreted from differentiating HL-60 cells following stimulation with PMA. In similar experiments with THP-1 cells, we observed that the PC was secreted to a low degree from untreated cells and to a high degree from cells treated continuously with PMA.

Inhibition of Cell Multiplication by the PC. Incubation of HL-60 cells with purified PC resulted in a time- and dose-dependent decrease in the rate of HL-60 cell multiplication. Inhibition of multiplication was observed even after 2 days of treatment with the PC at a concentration as low as 5 nM. This inhibitory effect was also evident in THP-1 cells, as well as in other cell lines (Table 1).

In contrast to its ability an inhibit cell multiplication, the PC at concentrations of up to 7.5 nM did not induce a mature phenotype in the HL-60 cells (i.e., the treated cells did not exhibit markers associated with myelomonocytic cell differentiation, such as reactivity with OKM1 mAb, reduction of nitrotetrazolium dye, or staining for nonspecific esterase activity).

Table 1 PC concentration required to produce 50% inhibition of cell multiplication

Cell Type	Concentration[a] (nM)
HL-60 promyelocytic leukemia	8
THP-1 monocytic leukemia	5
CEM T-lymphocytic	8
PN-3 malignant melanoma	5
U-373 glioblastoma	10
MCF-7 mammary adenocarcinoma	15
Y-79 retinoblastoma	15
IMR-70 lung fibroblast (nontransformed)	15

[a] Cells were seeded into 24-well tissue culture plates at 3×10^5 cells in 1 ml of medium and treated with the PC immediately after plating. After four days, the number of cells was determined by hemocytometer counting. (From Murao et al. 1990)

Similarly, treatment of CEM T-lymphoblastic leukemia cells with the purified PC reduced cell growth but did not alter reactivity with OKT3, OKT4, or OKT8 mAb's, which detect maturation antigens in this cell line (Murao et al. 1990).

To study the PC mode of action, we treated HL-60 cells with the complex and measured thymidine, uridine, methionine, and phosphate incorporation. The PC caused a dose-dependent inhibition of cellular incorporation of [^{32}P]phosphate, [^{3}H]thymidine, [^{3}H]uridine, and [^{35}S]methionine, with the assay of [^{32}P]phosphate incorporation being the most sensitive and [^{3}H]uridine incorporation being the least sensitive (Fig. 6). The inhibitory effect of the PC on phosphate incorporation was detected at less than 1 nM and at much higher concentrations by [^{3}H]thymidine, [^{35}S]methionine, and [^{3}H]uridine. The slopes of the incorporation profiles were linear for at least 3 h after addition of the radiolabeled precursor.

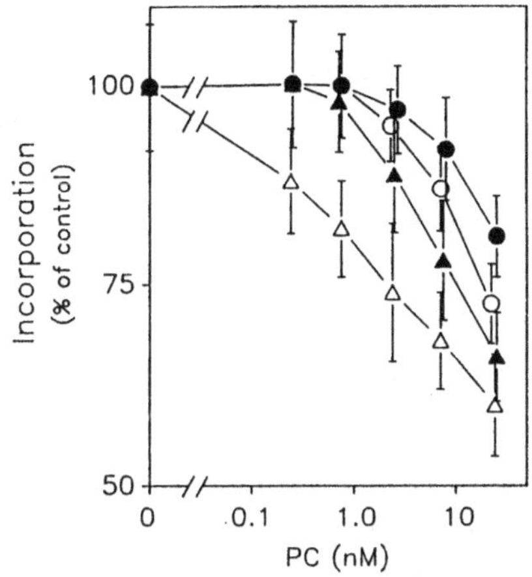

Fig. 6. Dose response for precursor incorporation in PC-treated HL-60 cells. Cells were seeded into 96-well tissue culture plates (3×10^6 cells/ml) in 200 μl of medium and treated with the purified PC immediately after plating. Incorporation of [^{32}P]phosphate (△), [^{3}H]thymidine (▲), [^{3}H]uridine (●), or [^{35}S]methionine (O) after PC treatment is presented as a percentage of the incorporation into the untreated cells. (From Murao et al. 1990)

We also tested the effect of the constituent MRP-8 and MRP-14 proteins on [^{32}P]phosphate incorporation. Each of these proteins inhibited the incorporation of [^{32}P]phosphate in a dose-dependent manner. However, on a dose basis, these two proteins were approximately 10-fold less effective than the PC itself. Since [^{32}P]phosphate incorporation was the most sensitive measure of PC-mediated inhibition, we analyzed the effect of the PC addition on the uptake of [^{32}P]phosphate into HL-60 cells in order to determine the contribution of uptake in the PC-mediated inhibition of [^{32}P]phosphate incorporation. The uptake of [^{32}P]phosphate increased nonlinearly over a 60-min incubation period; however, no significant decrease in [^{32}P]phosphate uptake was observed in HL-60 cells treated with the PC at doses up to 15 nM.

Possible Mode of Action of the PC. The mechanisms by which this complex mediates the reduction in cell multiplication are not clear; however, addition of the purified PC to cells in culture affects the incorporation of a number of metabolic precursors. The incorporation of

[32P]phosphate was the assay most sensitive to PC-mediated inhibition. A significant decrease of [32P]phosphate incorporation following doses of 1 nM or less of the PC preceded the inhibition of cell growth, which occurred at doses of 5–10 nM. The effect of the PC on [32P]phosphate incorporation was not attributable to changes in [32P]phosphate uptake nor to a reduction in the amount incorporated into specific cellular proteins, suggesting that the PC might affect an early event of phosphate metabolism. It is possible that this event may cause inhibition of cell multiplication directly or may cause it indirectly by reducing the incorporation of precursors into DNA, RNA, or protein.

In addition, the ability of the PC to inhibit protein kinase I and II (Murao et al. 1989) may have a role in the PC-mediated inhibition of cell multiplication. In particular, casein kinase II has been implicated in the phosphorylation of a variety of critical cellular proteins and in the processes of cell multiplication, transformation, and differentiation (Murao et al. 1989). It is thus possible that the multiplication-inhibitory effect of the PC is mediated by its ability to inhibit casein kinase II. However, the concentrations of the PC required to inhibit cell multiplication by 50% (5–15 nM) (Murao et al. 1990) are considerably lower than the K_i values for casein kinase II (200 nM) (Murao et al. 1989). However, this difference in PC concentrations does not eliminate a more specific inhibitory action of the complex on the enzyme *in vivo* or on an as-yet-unknown protein kinase (perhaps related to casein kinase II) that is critical for cell multiplication.

Regardless of these possibilities, the expression of the PC during differentiation and its multiplication-inhibitory effect in different cell types suggest that it plays a role in the physiological functions of myelomonocytic cells in addition to its possible role as a mediator of cell multiplication inhibition during terminal differentiation of these cells.

Further studies on the involvement of PKC β in the signal transduction processes that result in differentiation induction and of the PC in regulating cell replication during terminal differentiation of myelomonocytic cells will provide further insight into these critical cellular events. Moreover, this type of information will lead to a greater understanding of the processes leading to terminal differentiation and the aberrations in these processes that result in malignant transformation.

Acknowledgment

This work was supported by the U. S. Department of Energy, Office of Health and Environmental Research, under Contract no. W-31-109-ENG-38.

References

Anderson NL, Gemmell MA, Coussens PM, Murao S, and Huberman E (1985) Specific protein phosphorylation in human promyelocytic HL-60 leukemia cells susceptible or resistant to induction of cell differentiation by phorbol-12-myristate-13-acetate. Cancer Res., *45*: 4955–4962.

Blumberg PM, Complexities of the protein kinase C pathway. Molec. Carcinogenesis, *4*: 339–344.

Chiu R, Angel P, and Karin M (1989) Jun-B differs in its biological properties from, and is a negative regulator of, c-Jun. Cell, *59*: 979–986.

Dorin J, Novak M, Hill RE, Brock DJH, and Secher DS (1987) A clue to the basic defect in cystic fibrosis from cloning the CF antigen gene. Nature, *326*: 614–617.

Foon KA, Schroff RW, and Gale RP (1982) Surface markers on leukemia and lymphoma cells: recent advances. Blood, *60*: 1–19.

Halazonetis TD, Georgopoulos K, Greenberg ME, and Leder P (1988) c-Jun dimerizes with itself and with c-Fos, forming complexes of different DNA binding affinities. Cell, *55*: 917–924.

Hashimoto K, Kishimoto A, Aihara H, Yasuda I, Mikawa K, and Nishizuka Y (1990) Protein kinase C during differentiation of human promyelocytic leukemia cell line, HL-60. FEBS Lett., *263*: 31–34.

Homan EC, Jensen DE, and Sando JJ (1991) Protein kinase C isozyme expression in phorbol ester-sensitive and -resistant EL4 thymoma cells. J. Biol. Chem., *266*: 5676–5681.

Homma Y, Gemmell MA, and Huberman E (1988) Protein kinase C activities with different characteristics, including substrate specificity, from two human HL-60 leukemia cell variants. Cancer Res., *48*: 2744–2748.

Homma Y, Henning-Chubb C B, and Huberman E (1986) Translocation of protein kinase C in human leukemia cells susceptible or resistant to differentiation induced by phorbol 12-myristate 13-acetate. Proc. Natl. Acad. Sci. USA, *83*: 7316–7319.

Huang K-P (1990) Role of protein kinase C in cellular regulation. Biofactors, *2*: 171–178.

Jensen DE, Frankis RC Jr, and Sando JJ (1991) Defective induction of Jun and Fos-related proteins in phorbol ester-resistant EL4 mouse thymoma cells. Oncogene, *6*: 1219–1225.

Murao S, Collart F, and Huberman E (1990) A protein complex expressed during terminal differentiation of monomyelocytic cells is an inhibitor of cell growth. Cell Growth & Differen., *1*: 447–454.

Murao S, Collart FR, and Huberman E (1989) A protein containing the cystic fibrosis antigen is an inhibitor of protein kinases. J. Biol. Chem., *264*: 8356–8360.

Murao S, Epstein AL, Clevenger CV, and Huberman E (1985) Expression of maturation-specific nuclear antigens in differentiating human myeloid leukemia cells. Cancer Res., *45*: 791–795.

Nishizuka Y (1988) The molecular heterogeneity of protein kinase C and its implications for cellular regulation. Nature, *334*: 661–665.

Odink K, Cerletti N, Bruggen J, Clerc RG, Tarcsay L, Zwadlo G, Gehards G, Schlegel R, and Sorg C (1987) Two calcium binding proteins in infiltrate macrophages of rheumatoid arthritis. Nature, *330*: 80–82.

Schütte J, Viallet J, Nau M, Segal S, Fedorko J, and Minna J (1989) *jun*-B inhibits and c-*fos* stimulates the transforming and *trans*-activating activities of c-*jun*. Cell, *59*: 987–997.

Szabo E, Preis LH, Brown PH, and Birrer MJ (1991) The role of *jun* and *fos* gene family members in 12-*O*-tetradecanoylphorbol-13-acetate-induced hemopoietic differentiation. Cell Growth & Differ., *2*: 475–482.

Tonetti DA, Horio M, Collart FR, and Huberman E (1992) The protein kinase β gene expression is associated with susceptibility of human promyelocytic leukemia cells to phorbol ester-induced differentiation. Cell Growth & Differ., *3*: 739–745.

SIGNAL TRANSDUCTION THROUGH THE EPSTEIN-BARR VIRUS RECEPTOR IN HUMAN B LYMPHOCYTES

R. FRADE
Immunochimie des Régulations Cellulaires et des Interactions Virales
INSERM U.354
Hôpital Saint-Antoine
75012 Paris
France

INTRODUCTION

The Epstein-Barr virus receptor (EBV/C3dR, CR2 or CD21), originally isolated from the human B lymphoma cell line, Raji and purified by affinity chromatography on sepharose bound C3b/C3bi, was identified as gp140, a 140 kDa C3 binding glycoprotein (Barel et al., 1981).

The C3 binding site recognized by EBV/C3dR is the LYNVEA sequence (Lambris et al., 1985) expressed on C3d, C3dg, iC3b and to a lesser extent on C3b (Frade et al., 1985a; Frade and Strominger, 1980). The demonstration that C3dR (CR2) is also the Epstein-Barr virus (EBV) receptor was established by using specific polyclonal anti-gp140, prepared against the highly purified C3 receptor (Frade et al., 1984) and monoclonal anti-CR2, prepared randomly against cell surface antigens (Frade, 1986). Polyclonal anti-gp140 antibodies inhibited specifically EBV binding on Raji cell surface and prevented transformation of normal B lymphocytes by EBV (Frade et al., 1985b). EBV binds to the solubilized gp140 molecule immobilized on HB-5, a monoclonal anti-EBV/C3dR antibody (Fingeroth et al., 1984). The binding of EBV on EBV/C3dR is mediated by gp350/220 viral capsid protein (Nemerow et al., 1987). Presence of three sequence homologies between human C3d and gp350/220, the viral capsid protein and inhibition of C3d and EBV binding on EBV/C3dR by OKB-7, a monoclonal anti-EBV/C3dR antibody (Nemerow et al.,

NATO ASI Series, Vol. H 75
Cancer Therapy
Edited by N. D'Alessandro, E. Mihich, L. Rausa,
H. Tapiero, and T. R. Tritton

1987) suggested that EBV/C3dR expressed identical binding site for its two distinct extracellular ligands. However, despite these sequence homologies we have demonstrated, using monoclonal anti-EBV/C3dR antibodies and anti-idiotypic anti-EBV/C3dR antibodies prepared by using the highly purified receptor as original immunogen, that EBV/C3dR carries distinct binding sites for EBV and C3d (Barel et al., 1988). Presence of distinct binding sites on EBV/C3dR for C3d and EBV has been recently confirmed by others using human-mouse chimeras of EBV/C3dR (Molina et al., 1991; Martin et al., 1991).

EBV/C3dR cDNA sequence was established and deduced amino acid sequence supported that this receptor is characterized by an extracellular domain of 15 to 16 short consensus repeats of 60-70 amino acids, a transmembrane hydrophobic domain of 25 amino acids and a C-terminal cytoplasmic domain of 34 amino acids (Wong et al., 1985; Moore et al., 1987; Fujisaku et al., 1989). EBV/C3dR gene is located on band q32 of human chromosome 1, as genes of CR1, decay accelerating factor and C4-binding protein, on a 750-kb genomic restriction fragment, the Regulators of Complement Activation (RCA) genetic locus (Rey-Campos et al., 1988).

EBV/C3DR IS INVOLVED IN REGULATION OF HUMAN B LYMPHOCYTES

EBV/C3dR is involved in human B lymphocyte activation. Cross-linking of EBV/C3dR by specific ligands such as polyclonal anti-EBV/C3dR $F(ab')_2$ fragments, particle-bound C3d, a monoclonal anti-EBV/C3dR antibody (MoAb) OKB-7, or UV-inactivated EBV particles led to the enhancement of B cell proliferation in synergy with T cell factors, such as BCGF (Frade et al., 1985c; Masucci et al., 1987; Frade et al., 1986). "In vitro" activation of either human peripheral blood B lymphocytes by cell surface ligands, such as anti-IgM or SAC (Barel et al., 1986) or human tonsil B lymphocytes by the PKC activator, PMA (Changelian et al., 1986), induced "in vivo" phosphorylation of EBV/C3dR. EBV/C3dR isolated from plasma membrane was "in vitro" phosphorylated on serine and tyrosine residues (Delcayre et al., 1987).

Signal transduction through EBV/C3dR was analyzed by developping distinct and complementary analysis of: 1) its interaction at the cell surface with other cell surface antigens. Matsumoto et al.. (1991) demonstrated that

EBV/C3dR could interact with CD19, another cell surface antigen involved directly in human B lymphocyte activation; 2) its intracellular localization; 3) its interactions with intracellular components.

INTRACELLULAR INTERACTIONS OF EBV/C3DR

1) Intracellular localization of EBV/C3dR: Intracellular localization of EBV/C3dR was analyzed by preparing highly purified sub-cellular fractions of Raji cells and by following "in vitro" phosphorylation of this solubilized receptor. We demonstrated presence of phosphorylated EBV/C3dR in purified nuclei of Raji cells. Solubilized nuclear EBV/C3dR reacted with a nuclear ribonucleoprotein, p120 RNP, phosphorylated on serine residues. Cell-free phosphorylation of p120 RNP was dependent on the presence of solubilized EBV/C3dR (Delcayre et al., 1987; Delcayre et al., 1989).

Recently, we have analyzed the nuclear localization of EBV/C3dR by electron microscope immunochemistry of thin sections of Raji cells and by using monoclonal anti-EBV/C3dR prepared against the highly purified receptor. Anti-EBV/C3dR mAb immunogold labeling of thin sections of Raji cells identified EBV/C3dR at the nuclear surface and also within the nucleus. Nuclear envelope associated EBV/C3dR was localized mainly at nuclear pores. Within the nucleus, EBV/C3dR was associated with ribonucleoprotein (RNP) interchromatin fibrils. These data are in good agreement with our previous demonstration that solubilized EBV/C3dR interacts with a nuclear p120 RNP. EBV/C3dR expressed on the surface of purified nuclei interacted, as the EBV/C3dR expressed on the cell surface, with soluble and particle-bound C3bi/C3d (Gauffre et al., 1992).

2) Intracellular interactions of EBV/C3dR: Interactions between EBV/C3dR and other intracellular components were recently demonstrated. For this purpose, we prepared polyclonal anti-idiotypic anti-EBV/C3dR antibodies (Ab2), raised against $F(ab')_2$ fragments of polyclonal anti-EBV/C3dR antibodies (Ab1). The latter were originally raised against the whole solubilized and highly purified gp140 molecule. We postulated that this Ab2 carried specificities which mimicked extracellular or intracellular domains of

EBV/C3dR, reacting with extracellular or intracellular ligands, respectively (Barel et al., 1988).

As already mentionned above, we demonstrated that anti-idiotypic anti-EBV/C3dR antibodies (Ab2) carried specificities which mimicked extracellular domains reacting with its ligands, as human C3d and EBV, supporting presence of distinct extracellular binding sites on EBV/C3dR
(Barel et al., 1988).

In addition, this Ab2 led us to demonstrate that EBV/C3dR interacts, in the human B lymphoma cell line Raji, with the p53 cellular anti-oncogene-encoded phosphoprotein. P53 was not detected in normal B lymphocytes. Direct protein-protein interaction between EBV/C3dR and p53 was also demonstrated (Barel et al., 1989). In addition, we demonstrated that EBV/C3dR interacts, in normal B lymphocytes, directly with p68, an intracellular calcium-binding protein, belonging to the annexin VI family. P68 is an intracellular component distinct from the p53 anti-oncoprotein and is not expressed in transformed cells (Barel et al., 1991). Protein-protein interaction between EBV/C3dR and p68 is calcium dependent. Recently, we demonstrated that the regulatory proteins p53 and p68 bind specifically and directly on two distinct sites on the intracellular carboxy-terminal domain of EBV/C3dR. These data demonstrated that, despite its short length of 34 amino acids, the intracytoplasmic C-terminal tail of CR2 allows direct protein-protein interactions on two distinct binding sites of the Epstein-Barr virus and C3d receptor, with the two intracellular regulatory components, p53 anti-oncoprotein involved in tumor progression or p68 calcium-binding protein involved in control of cell proliferation (Frade et al., 1992).

While the biological role of these interactions remains unknown, the properties of EBV/C3dR to react with these two intracellular regulatory components and with the nuclear p120 ribonucleoprotein and its localization at the cell surface but also in the nucleus of the Burkitt B lymphoma cells, lead us to suggest that EBV/C3dR may act as an intracellular factor regulating proliferation of normal or transformed lymphocytes.

ACKNOWLEDGMENTS

The author would like to acknowledge Dr. Monique BAREL for helpful discussions.

REFERENCES

Barel M, Charriaut C, Frade R (1981) Isolation and characterization of a C3b receptor-like molecule from membranes of a human B lymphoblastoid cell line (Raji). FEBS Letters 136: 111-114.

Barel M, Vazquez A, Charriaut C, Aufredou MT, Galanaud P, Frade R. (1986) Gp140, the C3d/EBV receptor (CR2) is phosphorylated upon in vitro activation of human peripheral B lymphocytes. FEBS Letters. 197: 353-356.

Barel M, Fiandino A, Delcayre AX, Lyamani F, Frade R (1988) Monoclonal and anti-idiotypic anti-EBV/C3d receptor antibodies detect two binding sites, one for EBV and one for C3d on gp 140, the EBV/C3dR expressed on human B lymphocytes. J. Immunol. 141: 1590-1595.

Barel M, Fiandino A, Lyamani L, Frade R (1989) Epstein-Barr virus/C3d receptor (EBV/C3dR) reacts with p53, a cellular anti-oncogene-encoded membrane phosphoprotein: detection by polyclonal anti-idiotypic anti-EBV/C3dR antibodies (Ab2). Proc. Natl. Acad. Sci. USA. 86: 10054-10058.

Barel M, Gauffre A, Lyamani F, Fiandino-Tirel A, Hermann J, Frade R. (1991) Intracellular interaction of EBV/C3d receptor (CR2) with p68, a calcium-binding protein present in normal but not in transformed B lymphocytes. J. Immunol. 147:1286-1291.

Changelian PS, Fearon DT (1986) Tissue-specific phosphorylation of complement receptors CR1 and CR2. J. Exp. Med. 163: 101-115.

Delcayre AX, Fiandino A, Barel M, Frade R (1987) Gp 140, the EBV/C3dR receptor (CR2) of human B lymphocytes is involved in cell-free phosphorylation of p120, a nuclear ribonucleoprotein. Eur. J. Immunol. 17: 1827-1833.

Delcayre AX, Fiandino A, Lyamani F, Barel M, Frade R (1989) Enhancement of Epstein-Barr virus/C3d receptor (EBV/C3dR or CR2) and nuclear p120 ribonucleoprotein phosphorylation by specific EBV/C3dR ligands in subcellular fractions of the human B lymphoma cell line, Raji. Biochem. Biophys. Res. Commun. 159: 1213-1220.

Fingeroth JD, Weis JJ, Tedder TF, Strominger JL, Biro PA, Fearon DT. (1984) Epstein-Barr virus receptor of human B lymphocytes is the C3d receptor CR2. Proc. Natl. Acad. Sci. USA 81: 4510-4514.

Frade R, Strominger J. (1980) Binding of soluble ^{125}I human C3b, the third component, to specific receptor in human cultured B lymphoblastoid cells: characterization of a low affinity interaction. J. Immunol., 125: 1332-1339.

Frade R, Barel M, Krikorian L, Charriaut C (1984) Analysis of gp 140, a C3b binding membrane component present on Raji cells: a comparison with H. Eur. J. Immunol 14: 542-548.

Frade R (1986) Structure and functions of gp 140, the EBV/C3dR receptor (CR2) of human B lymphocytes. Mol. Immunol. 23: 1249-1253.

Frade R, Myones B, Barel M, Krikorian L, Charriaut C, Ross G (1985a) Gp 140, a C3b binding membrane component of B lymphocytes is the C3d/C3dg receptor (CR2) and is distinct from the neutrophils C3dg receptor (CR4). Eur. J. Immunol., 15: 1192-1197.

Frade R, Barel M, Ehlin-Henriksson B, Klein G (1985b) Gp 140, the C3d receptor (CR2) of human B lymphocytes is also the EBV receptor. Proc. Natl. Acad.Sci. USA 82: 1490-1493.

Frade R, Crevon MC, Barel M, Vazquez A, Krikorian L, Charriaut C, Galanaud P (1985c) Enhancement of human B cell proliferation by an antibody to the C3d receptor, the gp 140 molecule. Eur. J. Immunol.15: 73-76.

Frade R., Gauffre A, Hermann J, Barel M. (1992) Epstein-Barr virus/C3d receptor (CR2) interacts by its intracytoplasmic carboxy-terminal domain and distinct binding sites with the p53 anti-oncoprotein and the p68 calcium-binding protein. J. Immunol. 149: in press.

Fujisaku A, Harley JB, Frank MB, Gruner BA, Frazier B, Holers VM (1989) Genomic organization and polymorphisms of the human C3d/Epstein Barr virus receptor. J. Biol. Chem. 264: 2118-2125.

Gauffre A, Viron A, Barel M, Hermann J, Puvion, Frade R. (1992) Nuclear localization of the Epstein-Barr virus/C3d receptor (CR2) in the human Burkitt B lymphoma cell, Raji. Mol. Immunol. 29: 1113-1120.

Lambris JD, Ganu VS, Hiran S, Muller-Eberhard HJ (1985) Mapping of the C3d receptor-binding site and neoantigenic site in the C3d domain of the third component of complement. Proc. Natl. Acad. Sci. USA. 82: 4235-4239.

Martin, D. R., A. Yuriev, K. R. Kalli, D. T. Fearon, J. M. Ahearn. (1991) Determination of the structural basis for selective binding of Epstein-Barr virus to human complement receptor type-2. J. Exp. Med. 174:1299-1308.

Masucci MG, Szigeti R, Ernberg I, Hu CP, Torsteinsdottir S, Frade R, Klein E (1987) Activation of B lymphocytes by Epstein-Barr virus/CR2 receptor interactions. Eur. J. Immunol. 17: 815-820.

Matsumoto AK, Kopicky-Burd J, Carter RH, Tuveson DA, Tedder TF, Fearon DT. (1991) Intersection of the complement and immune system: A signal transduction complex of the B Lymphocyte-containing complement receptor type 2 and CD19. J. Exp. Med. 173: 55-64.

Molina, H., C. Brennert, S. Jacobi, J. Gorka, J. C. Carel, T. Kinoshita, V. M. Holers. (1991) Analysis of Epstein-barr virus-binding sites on complement receptor 2 (CR2/CD21) using human-mouse chimeras and peptides. At least two distinct sites are necessary for ligand-receptor interaction. J. Biol. Chem. 266: 12173 -12779.

Moore MD, Cooper NR, Tack BF, Nemerow GR (1987) Molecular cloning of the cDNA encoding the Epstein-Barr virus /C3d receptor (CR2) of the human B lymphocyte. Proc. Natl. Acad. Sci USA. 84: 9194-9198.

Nemerow GR, Mold C, Schwend VK, Tollefson V, Cooper NR. (1987). Identification of gp 350 as the viral glycoprotein mediating attachment of Epstein-Barr virus to the EBV/C3d receptor of B cells: sequence homology of gp 350 and C3 complement fragment C3d. J. Virol., 61: 1416-1421.

Rey-Campos J, Rubinstein P, Rodriguez de Cordoba S. (1988) A physical map of the human regulator of complement activation gene cluster linking the complement genes CR1, CR2, DAF and C4Bp. J. Exp. Med. 167: 664-672.

Wong WW, Klickstein LB, Smith JA, Weis JH, Fearon DT (1985). Identification of a partial cDNA clone for the human receptor for complement fragments C3b/C4b. Proc. Natl. Acad. Sci. USA. 82: 7711-7716.

SIGNAL TRANSDUCTION MECHANISMS AS A TARGET FOR CANCER CHEMOTHERAPY

Thomas R. Tritton
Department of Pharmacology and
Vermont Cancer Center
University of Vermont College of Medicine
Burlington, VT 05405 USA

Mechanism of Action of Adriamycin

Adriamycin is one of the most useful drugs employed to treat human cancer and, because of this central importance, there has been considerable effort devoted to understanding its mechanism of action. As summarized in Table 1, several proposals have been offered to explain the anticancer activity of adriamycin and other members of the anthracycline family.

Table 1

Proposed Mechanisms of Action for Adriamycin

A. Intercalation of the drug into the DNA double helix
B. Generation of reactive oxygen through one-electron reduction
C. Generation of alkylating species through two-electron reduction
D. Stabilization of topoisomerase II/DNA cleavable complexes
E. Damage to the plasma membrane

There is evidence both for and against each of these proposals (Neidle, 1979; Bachur, *et al., 1979*; Moore, 1977; Tewey, *et al.*, 1984; Tritton & Hickman, 1985). To limit the length of this discussion, we will provide only a synopsis but recognize that this issue is not settled and remains under active investigation in several laboratories.

Intercalation (mechanism A) is historically important because it was the first proposal offered to explain adriamycin action, but since anthracyclines exist which have little or

NATO ASI Series, Vol. H 75
Cancer Therapy
Edited by N. D'Alessandro, E. Mihich, L. Rausa,
H. Tapiero, and T. R. Tritton

no affinity for DNA (e.g. AD 32 - (Pearlman *et al.*, 1986), and N,N-dibenzyladriamycin - (Acton, 1980)), it is now reasonably clear that intercalation per se is not a necessary and sufficient condition to bring about drug action. It has been argued that these compounds are metabolized to derivatives which do bind DNA. This is not universally observed, however, (Levin *et al.*, 1981; Gamba-Vitalo, *et al.*, 1987), so interaction with other sites like topoisomerases may be a more likely explanation for the cytotoxicity of AD-32 and related drugs (Ross, 1985; Potmesil, *et al.*, 1986).

Production of reactive oxygen (mechanism B) is attractive because adriamycin is capable of engaging in redox reactions to produce superoxide, peroxide and/or hydroxyl radicals; these species are known to be toxic. Arguing against this mechanism, however, are two facts: [a] certain hypoxic cells (*i.e.* low oxygen) are more susceptible to adriamycin cytotoxicity than oxygenated cells (Siegfried *et al.*, 1983*a*), and [b] there is no structure/activity correlation between cytotoxicity and production of oxygen radicals by anthracycline congeners (Burke *et al.*, 1987).

Generation of alkylating species (mechanism C) is also theoretically possible but ruled against by a lack of relationship between covalent attachment of the drug to DNA with either cytotoxicity or drug resistance (Siegfried *et al.* 1983*a*; Siegfried *et al.*, 1983*b*).

The role of topoisomerases in cytotoxic action (mechanism D) is under intense study for several anticancer drugs, including adriamycin (Liu, 1989). Our thinking is that this enzyme does play a key role, but probably not by itself *i.e.* other factors to be explained below are also participants in the cytotoxic process.

The list of membrane actions (mechanism E) of adriamycin is very extensive and includes a variety of effects on signal transduction systems like phosphoinositide metabolism, protein kinase C, EGF receptor action, Ca^{++} signalling, cyclic nucleotides, among others (reviewed in Tritton & Hickman, 1985; Tritton & Hickman, 1990). However, not all of the proposed mechanisms are mutually exclusive and a sequential or parallel combination of actions may explain the biologic activity of adriamycin.

Temperature Studies

Recent work by investigators active in the anthracycline field has sought to learn which target loci are essential for adriamycin-induced cytotoxicity, and how the disparate pathways may be linked to bring about cell death. In our laboratory we have found that temperature is a very useful variable in this regard. Even though adriamycin does gain access to the cell interior at low temperature, when we measured the temperature dependence of cell death we were surprised to find that at all temperatures below about 20°C there is no cytotoxic response to the presence of the drug. In fact, no matter how high the adriamycin concentration or how long the exposure, as long as the temperature is below the critical 20°C there is no untoward cellular response to the presence of this otherwise noxious agent. By itself this result is not definitive, but if one assumes that the critical target for adriamycin is inside the cell, then there should be a direct relationship between uptake and cell death (*i.e.* the more uptake, the more likely the cell will die). Clearly this relationship does not hold, so from the temperature results one is inclined to favor an extracellular (plasma membrane) target for drug action since this location would not be expected to depend on intracellular uptake.

If uptake does not govern cytotoxicity, some other factor must be operative since the drug is surely capable of causing cell death. Table 2 lists several variables we have examined in an attempt to rationalize the unusual temperature dependence of adriamycin's action (summarized from Lane, 1987; Vichi *et al.*, 1989). Without describing the details, one sees in the Table that neither uptake, nor metabolism, nor subcellular distribution, nor redox reactions can explain the temperature dependence of drug action.

Table 2

Variables examined to explain the temperature dependence of Adriamycin's ability to cause cell death.

Variable	*Correlation with cytotoxicity temperature profile*
uptake	no
metabolic conversion	no
subcellular distribution	no
redox reactions	no
membrane fluidity	yes
DNA damage	yes

One cellular property we find which does show a temperature dependence reminiscent of the cytotoxicity profile is the fluidity of the plasma membrane. The temperature dependence of adriamycin intrinsic fluorescence polarization when bound to the plasma membrane of L1210 cells clearly shows a discontinuity at about 20°C (Lane, 1987). Such behavior is generally taken as presumptive evidence for a membrane phase change (solid↔fluid transition) at the indicated temperature. Thus, the binding site(s) for drug on the cell surface do exhibit a structural change at the same temperature that the drug becomes incapable of causing cell death. This could be a coincidence, but we think it equally likely that the structure of the plasma membrane is involved in regulating the response to drug, and this was among the clues suggesting that signalling mechanisms could be important.

The Role of DNA Damage

There is considerable sentiment among cancer chemotherapists that DNA is a major target for adriamycin. The drug does bind to DNA by intercalation and does cause DNA damage, so we decided to examine the temperature dependence of DNA lesions to see if this property resembled the temperature dependence of cytotoxicity (Vichi *et al.*, 1989). Most laboratories measure adriamycin-induced DNA damage by alkaline elution. We found that the temperature dependence of formation for both single strand breaks (SSB) and DNA-protein crosslinks (DPC) is identical to the temperature dependence of cytotoxicity *i.e.* above 20°C there is DNA damage and it is accompanied by cell death, while below 20°C there is neither DNA damage nor cell death. Thus, these results strongly suggest a functional linkage between adriamycin's ability to derange DNA and its ability to kill the cell.

DNA is not damaged by adriamycin itself, but by the aberrant action of the enzyme topoisomerase II (Tewey, *et al.*, 1984). Malfunction of this enzyme yields both DNA breaks and stabilized DNA-topoisomerase complexes. Thus, another potential explanation for adriamycin's inability to kill cells at low temperature would be that topoisomerase II becomes inactive when cooled. We tested this possibility (Vichi, *et al.*, 1989) by measuring the ability of topoisomerase to act on two different substrates: knotted P4 DNA and superhelical pBR322 DNA. The results show that the enzyme is functional at low temperature (albeit with a reduced catalytic rate) and that adriamycin can still disrupt its ability to process DNA correctly. Consequently, altered topoisomerase II activity cannot explain the temperature dependence of DNA damage and cell death wrought by adriamycin.

We now have an apparent paradox: adriamycin cannot kill cells at low temperature nor provoke DNA damage, but the enzyme that causes this damage remains perfectly functional, even at 0°C. A way to rationalize this dilemma would be to postulate the existence of another cellular factor which regulates topoisomerase II activity. To assess this possibility we isolated cell nuclei, discarding all cytoplasmic and membrane components which might interact with topoisomerase II. The ensuing experimental results showed that adriamycin does cause DNA damage in isolated nuclei at 37°C and, unlike with whole cells, also at 0°C. The DNA damage in isolated nuclei at 0°C is both time- and dose-dependent as one would expect. Thus, purifying the nuclei does appear to eliminate a factor which causes the DNA to become unresponsive to damage by adriamycin in intact cells.

How can it be true both that nuclear DNA damage is required for cytotoxicity and that drug must interact with other cellular factors to cause cytotoxicity? **This situation implies**

a communication between the nucleus and the rest of the cell, and suggests that adriamycin disruption of signal transduction systems may provide the mechanism by which cell death is initiated with this chemotherapeutic agent. In fact, it has been shown (Posada, *et al.*, 1989*a*) that adriamycin interaction with the cell surface causes the set of responses indicated in Table 3. Thus, the protein kinase C pathway appears to be one candidate for a signal transduction pathway targeted by adriamycin, and the next section of this essay will present some consequences of this result.

Table 3

Changes in Signal Transduction Pathways Induced by Adriamycin

Changes
phosphoinositide turnover increased
diacylglycerol accumulation increased
protein kinase C activity increased

Serine/Threonine Phosphorylation of Topoisomerase II

Since adriamycin treatment of cells affects the activity of protein kinase C (Posada, *et al.*, 1989*a*; Posada *et al.*, 1989*b*), it is possible that the reverse will be true as well, *i.e.* modulation of the activity of PKC may in turn modulate the activity of adriamycin. To test this postulate, we turned to the phorbol ester TPA to alter PKC activity because this compound can both activate (by short treatment) and down regulate (by extended treatment) the catalytic activity of PKC. Table 4 summarizes results showing that activation (4-fold) of PKC with a 30 minute exposure to TPA causes a dose dependent increase in cytotoxicity. Conversely, when PKC activity is reduced (9-fold) by down regulation, there is less cytotoxicity. Most germane to the theme being developed here, TPA also induces corresponding alterations in the ability of topoisomerase II to damage DNA as measured by alkaline elution: activation of PKC causes more cytotoxicity and more DNA damage, and reduction of PKC activity has exactly the opposite effect. Thus, there appears to be a regulated connection between the activity of PKC, the ability of topoisomerase II to damage DNA, and the response of the cell to die.

Table 4

Effect of Phorbol Ester on Cellular Response to Adriamycin

TPA 30 Minutes:	PKC ↑	cell kill ↑	DNA damage ↑
TPA 24 hours	PKC ↓	cell kill ↓	DNA damage ↓

A mechanism that could explain these results is that PKC controls the activity of topoisomerase II by phosphorylation. The first question we addressed is: can DNA topoisomerase II from CEM human leukemic cells be activated by PKC purified from rabbit brain? To address this question topoisomerase II catalytic activity was measured by decatenation of kinetoplast DNA under different conditions. The results showed that 200 ng of nuclear extract is barely able to decatenate the substrate DNA under the conditions chosen unless PKC and its activating cofactors (Ca^{++}, phosphatidylserine) are also present. Appropriate controls have been done to show that PKC alone does not catalyze DNA decatenation, and that the activation of topoisomerase is directly related to the amount of added PKC. Thus, PKC can activate topoisomerase II, and this is the first such demonstration in human cells. We have also demonstrated by immunoprecipitation of ^{32}P labelled cells followed by SDS-PAGE that topoisomerase II is phosphorylated in intact CEM cells, and that this phosphorylation is reduced in CEM/VM-1, an altered topoisomerase MDR variant; Ganapathi's laboratory has also reported lowered topoisomerase phosphorylation in MDR cells (Ganapathi, *et al.*, 1991). The correspondence between phosphorylation and activity awaits further experimentation.

It should be stressed that our results to date do not prove that PKC is the *in vivo* regulator of topoisomerase, only that addition of the kinase modulates topoisomerase activity. There could be several linked phosphorylation reactions, and other responsible kinases also involved. Casein kinase II has also been reported to activate topoisomerase II in Drosophila (Ackerman, *et al.*, 1985; Ackerman, *et al.*, 1988). Heparin is a CKII inhibitor so we tested its ability to inhibit the activation of topoisomerase II with and without added PKC. No effect was demonstrable so we have tentatively concluded that CKII is not involved in topoisomerase II activation in our human cellular system.

Signal Transduction Mechanisms in Adriamycin Resistance

If signal transduction mechanisms are involved in the mechanism of *action* of adriamycin, it is also possible that these same mechanisms might be implicated in the mechanism of *resistance* to this drug. Adriamycin is a member of the family of agents in the multidrug resistant (MDR) class, so we have measured protein kinase C activity in sensitive and various multidrug resistant cells. Previous workers in several laboratories, including our own, have demonstrated a consistent elevation of PKC activity in classic P-glycoprotein mediated MDR (Posada *et al.*, 1989*a*; Posada *et al.*, 1989*b*; Fine *et al.*, 1988; O'Brian *et al.*, 1989; Aquino, *et al.*, 1990; Chambers, *et al.*, 1990; Lee, *et al.*, 1992). Presumably, an important substrate for PKC in this instance is P-gp itself, and we and others have published evidence in support of this notion (Posada *et al.*, 1989*a*; Chambers *et al.*, 1990). We have also examined the subcellular distribution of PKC activity in cytosol, membrane, and nuclear fractions from CEM sensitive, classic MDR, and AT-MDR cells. The significant conclusions are as follows: [1] the membrane activity of PKC is invariant among the three cell lines; [2] cytosolic PKC activity is elevated in Pgp MDR cells; [3] nuclear PKC activity is elevated in the AT-MDR line. This latter finding is particularly intriguing since it has not been reported previously by others. The basis for resistance in AT-MDR cell lines is thought to lie in altered topoisomerases, so the presence of modulated PKC activity in the nucleus could underlie the alteration in control of topoisomerase function.

We have also found in both mouse Sarcoma 180 and human KB cells selected for the multiple drug resistance (MDR) phenotype, there is an elevation in the steady-state mRNA level of *c-fos* There is no detectable gene amplification for *c-fos*, nor is there any significant change in the rate of mRNA transcription or degradation, suggesting that other factors are responsible for the increased expression level in resistance. Cells selected for resistance to methotrexate, a drug not in the MDR group, do not have an increase in *c-fos* mRNA expression. When drug sensitive cells are exposed for 30 minutes to an ED_{50} concentration of vinblastine, adriamycin, colchicine, or VP-16, but not to methotrexate or cisplatin, there is a three- to six-fold induction in the level of *c-fos* message. Since the former drugs are members of the multidrug resistance class and the latter are not, the results are consistent with the hypothesis that induction of *c-fos* by low levels of cytotoxic drugs may be an early event in the acquisition of the MDR phenotype. If this were the case then *c-fos* would be expected to act in concert with *c-jun* to control transcription by binding to a specific DNA regulatory site. Consistent with this explanation is the existence of an AP-1 sequence in the promotor region for the P-glycoprotein gene (*mdr1*), as well as the fact that *c-jun* is also overexpressed in MDR cells.

Table 5

Alterations in Signal Transduction and Transcription Factors in MDR

PKC activity ↑

PKC quantity ↑

PKC translocation by TPA ↓

PKC subcellular distribution is altered

c-fos ↑

c-jun ↑

If *c-fos* levels do in fact play a regulatory role over a cell's ability to acquire the MDR phenotype, then raising the level of *c-fos* should enhance the competence of a cell to become resistant. Consequently, we have transfected S180 cells with *c-fos* sequences and subjected the transfectants to selection for adriamycin resistance. The two most important conclusions are summarized in Table 6. First, cells that overexpress *c-fos* reach a higher level of adriamycin resistance, and do so more rapidly than cells with basal levels of *c-fos* when subjected to classic selection procedures. Second, the introduction of higher than normal levels of *c-fos* confers a resistance phenotype on the cells even in the absence of selective pressure. Thus, transfected clones have intrinsic resistance to adriamycin from two to ten fold, and also exhibit a full multidrug resistance phenotype even though no selection and no elevation of P-glycoprotein has occurred. The reason for this is unclear, but may derive from the fact that the elevated level of a transcription factor (*c-fos*) promotes the active transcription of a variety of genes, some of which (*e.g.* glutathione-S-transferase, cytochrome P450) might contribute to a resistance trait.

Table 6

Consequences of c-fos transfection of drug sensitive cells

multidrug resistance phenotype induced
higher frequency selection for P-glycoprotein variants

References

Ackerman P, Glover CVC, and Osheroff N (1985) Phosphorylation of DNA topoisomerase II by casein kinase II: modulation of eukaryotic topoisomerase II activity in vitro. Proc Natl Acad Sci 82:3164-3168

Ackerman P, Glover CVC, and Osheroff N (1988) Phosphorylation of DNA topoisomerase II in vivo and in total homogenates of drosophila kc cells. J Biol Chem 263:12653-12660

Acton EM (1980) N-alkylation of anthracyclines. In: Crooke ST and Reich SD (eds) Anthracyclines: Current Status and New Developments, Academic Press, New York, p 15

Aquino A, Warren BS, Omichinski J, Hartman KD and Glazer RI (1990) Protein kinase C_γ is present in adriamycin resistant HL-60 cells. Biochem Biophys Res Comm 166:723-728

Bachur NR, Gordon SL, Gee MV and Kon H (1979) NADPH Cytochrom P-450 reductase activation of quinone anticancer agents to free radicals. Proc Nat Acad Sci 76:954-957

Burke T, Pritsos C, Sartorelli AC and Tritton TR (1987) The structural basis for anthracycline antibiotic stimulation of oxygen consumption by HL-60 cells and mitochondria. Cancer Biochem Biophys 9:245-255

Chambers TC, McAvey EM, Jacobs JW, and Iom G (1990) Protein kinase C phosphorylates P-glycoprotein in multidrug resistant tumor KB carcinous cells. J Biol Chem 265:7679-7687

Fine RL, Patel J and Chabner BA (1988) Phorbal esters induce multidrug resistance in human breast cancer cells. Proc Nat Acad Sci 85:582-586

Gamba-Vitalo C, Blair OC, Tritton TR, Lane P, Carbone R and Sartorelli AC (1987) Cytotoxicity and differentiating actions of adriamycin in WEHI-3B D+ leukemia cells. Leuk R:1:188-197

Ganapathi R, Kamath N, Constantinov A, Grabowski D, Ford J, and Anderson A (1991) Effect of the calmodulin inhibitor trifluoperazine on phosphorylation of P-glycoprotein and topoisomerase II: Relationship to modulation of subcellular distribution, DNA damage, and cytotoxicity of doxorubicin in multidrug resistant L1210 mouse leukemia cells. Biochem Pharm 41:R21-R26

Lane P, Vichi P, Bain DL, and Tritton TR (1987) Temperature dependence studies of Adriamycin uptake and cytotoxicity. Cancer Res 47:4038-4042

Lee SA, Karaszkiewicz JW, and Anderson WB (1992) Elevated level of nuclear protein kinase C in multidrug resistant MCT-7 human breast carcinous cells. Cancer Res 52:3750-3759

Levin M, Silber R, Israel M, Goldfeder A and Khetarpal VK (1981) Protein associated DNA breaks and DNA-protein cross links caused by DNA nonbinding derivatives of adriamycin in L1210 cells. Cancer Res 41:1006-1010

Liu LF (1989) DNA topoisomerase poisons as antitumor drugs. Ann Rev Biochem 58:351-375

Moore WH (1977) Bioactivation as a model for drug design: bioreductive alkylation. Science 197:527-532

Neidle S (1979) The molecular basis for the action of some DNA-binding drugs. Prog Med Chem 16:151-220

O'Brian CA, Fan D, Ward NE, Scid C, and Fidler IJ (1989) Level of protein kinase C activity correlates directly with resistance to adriamycin in murine fibrosarcoma cells. Febs Lett 246:78-82

Pearlman LF, Chuang RY, Israel M and Simpkins H (1986) Interaction of three second-generation anthracyclines with polynucleotides, RNA, DNA, and nucleosomes. Cancer Res 46:341-346

Posada JA, McKeegan EM, Worthington KR, Morin NJ, Jaken S, and Tritton TR (1989*a*) Human multidrug resistant KB cells overexpress protein kinase C: involvement in drug resistance. Cancer Comm 1:285-292

Posada JA, Vichi P, and Tritton TR (1989*b*) Protein kinase C in adriamycin action and resistance in mouse S180 cells. Cancer Res 49:6634-6639

Potmesil M, Israel M, Kirschenbaum S, Sweatman TW and Siber R (1986) Kinetics of DNA protein-associated breaks induced by N-acyl and N-alkylanthracyclines as it relates to drug metabolism. Proc Amer Assoc Cancer Res 27:247

Ross W (1985) DNA topoisomerases as targets for cancer therapy. Biochem Pharm 34:4191-4195

Siegfried JM, Kennedy KA, Sartorelli AC and Tritton TR (1983*a*) Effect of anthracyclines on oxygenated and hypoxic tumor cells. Cancer Res 43:54-59

Siegfried JM, Sartorelli AC and Tritton TR (1983*b*) Comparison of anthracycline concentrations in S180 cell lines of varying sensitivity. Eur J Cancer Clin Oncol 19:1133-1141

Tewey KM, Rowe TC, Yang L, Halligan BD and Liu LF (1984) ADR-induced DNA damage mediated by mammalian DNA TOPO II. Science 226:466-468

Tritton TR (1991) Cell surface actions of anthracyclines. Pharm and Therap 49:293-309

Tritton TR and Hickman JA (1985) Cell membranes as a chemotherapeutic target. In: Muggia FM (ed) Experimental and Clinical Progress in Cancer Chemotherapy, Martinus Nijoff, Boston, p 82

Tritton TR and Hickman JA (1990) How to kill cancer cells: membranes and cell signals as targets for cancer chemotherapy. Cancer Cells 2:95-105

Vichi P, Robison S, and Tritton TR (1989) Temperature dependence of adriamycin-induced DNA damage in L1210 cells. Cancer Res 49:5575-5580

THERAPEUTIC APPROACHES FOR COLON CANCER BASED ON TRANSCRIPTIONAL REGULATION OF SPECIFIC GROWTH FACTORS

M.G. Brattain* and K.M Mulder+
*Department of Biochemistry and Molecular Biology, Medical College of Ohio, P.O. Box 10008, Toledo, OH 43699-0008
+Department of Pharmacology, Pennsylvania State University College of Medicine, 500 University Drive, Hershey, PA 17033

INTRODUCTION

Colon cancer, like many other solid tumors, is poorly responsive to chemotherapeutic approaches in terms of cure rates and increased survival time. As a consequence, this tumor has been the subject of extensive investigations seeking alternative or perhaps supplemental therapies based on biochemical, biological or immunological properties of colon cancer cells which are not shared by normal cells.

Our work has been based on the widespread assumption that control of growth and the cell cycle in normal cells is disrupted in malignant cells. Identification of the differences in growth control between normal and malignant cells could lead to the exploitation of specific growth related therapeutic targets. To this end we have concentrated our efforts on endogenously produced growth factors of colon cancer which are either autostimulatory or autoinhibitory to proliferation and DNA synthesis. Of particular interest in this regard were the contributions of these autocrine factors to the establishment of highly progressed cellular phenotypes which are independent of the need for exogenous growth factors and, therefore, independent of the growth regulatory signals which are operative in the normal cell environment to which normal cells are generally responsive.

During the early 1980's we developed a model system of a bank of human colon carcinoma cell lines which has been conducive for the identification of growth related targets. A large bank of cell lines was established from approximately 25 specimens of primary tumors. The individual cell lines in this bank reflected the heterogeneity of cell types associated with this disease in people (Brattain, et al., 1984; Chantret et al., 1988; Mulder and Brattain, 1989a). Two phenotypes were

NATO ASI Series, Vol. H 75
Cancer Therapy
Edited by N. D'Alessandro, E. Mihich, L. Rausa, H. Tapiero, and T. R. Tritton

of particular interest in terms of growth regulatory studies. The first phenotype designated Group I, consisted of highly aggressive cell lines which formed poorly differentiated tumors in athymic mice at low inocula while the other phenotype, designated Group III, consisted of cell lines which required very high inocula for sporadic tumor formation. When tumors were formed by Group III cells they showed a high degree of differentiation. The properties of these 2 phenotypes which are continuously maintained in fully defined tissue culture medium are summarized in Table 1 (Boyd et al., 1988; Wan et al., 1988). Most importantly for growth regulatory studies, Group I cells are completely growth factor independent with respect to DNA synthesis while Group III cells require exogenous growth factors for optimal DNA synthesis (Mulder and Brattain, 1989b; Mulder et al., 1990a,b).

TABLE 1: BIOLOGICAL PROPERTIES OF HUMAN COLON CARCINOMA CELLS

GROUP 1	GROUP III
Highly tumorigenic	Poorly tumorigenic
Poorly differentiated xenografts	Well-differentiated xenografts
No tight junctions	Tight junctions
No done formation in culture	Form domes in culture
Growth factor independent	Growth factor dependent
Express villin	Express villin

The purpose of this manuscript is to describe growth regulatory targets for which we have obtained evidence that they might be therapeutically exploitable and to discuss new approaches for anti-cancer drug identification based on transcriptional control of these identified growth regulatory factors. Evidence for the autocrine role of transforming growth factor α (TGFα) in the generation of the growth regulatory independent phenotype will be summarized. Moreover, a particularly pathologic form of the TGFα autocrine loop which is not susceptible to extracellularly directed agents such as antibodies will be shown to be responsible for growth regulatory independence through the transcriptional control of its own expression as well as the control of the expression of other stimulatory autocrine factors. Finally, evidence that growth regulatory independence emerges as a direct result of the breakdown of negative cell cycle control by transforming growth factor β (TGFβ) will be described.

TRANSFORMING GROWTH FACTOR α

As originally stated the autocrine hypothesis was generated to account for the relative independence of transformed cells from exogenous growth factors and thus their ability to control their own proliferation with minimal interference by normal growth controls (Sporn and Todaro, 1980). TGFα was identified as the prototype autocrine growth factor in both virally transformed cells and human tumor cell lines. TGFα which is a 50 amino acid product released proteolytically from a membrane bound pro-TGFα form interacts with the epidermal growth factor receptor (EGF_r) to induce DNA synthesis and cell proliferation (review, Massagué, 1990). Ligand binding to the EGF_r takes place at the cell surface and, therefore, neutralizing antibodies to either the EGF_r or TGFα can block receptor activation and thus inhibit DNA synthesis and proliferation in cells dependent upon autocrine TGFα for optimal growth (Gill et al., 1984). Initially it was thought that autocrine TGFα loops were restricted to cancer cells and, indeed, it has been shown that solid cancers in general show higher expression of TGFα and EGF_r than their normal counterpart tissues (Derynck et al., 1987). However, in recent years it has become clear that TGFα acts as an autocrine factor in normal cell types (Bates et al., 1989).

Variations of the classical autocrine loop have been identified or created by appropriate engineering of growth factor ligands. For example, cells in which a mutated TGFα was expressed which could not be proteolytically cleaved from the cell surface were shown to activate cell surface EGF_r in a second cell type which does not express TGFα (Wong et al., 1989; Brachmann et al., 1989). This type of autocrine loop has also been shown to occur natively in HT1080 cells which apparently lack the proteolytic enzyme(s) necessary for TGFα processing (Ankesaria et al., 1990).

Several growth factors have been implicated as forming intracellular autocrine loops (review Browder et al., 1989). For example, overexpression of interleukin-3 (IL-3) and platelet derived growth factor (PDGF) have been shown to lead to intracellular activation of the receptors for these ligands based on the inability of antibodies to interfere with the growth of the target cells. Moreover, modification of the ligands to express the SEKDEL golgi and endoplasmic reticulum anchor can also lead to receptor activation as indicated by the transformation of targeted cells. The intracellular localization of this type of autocrine loop can lead to constitutive receptor activation which is inaccessible to environmental control. Recently, we have identified a native TGFα intracellular autocrine loop which we refer to as a TGFα intracrine loop.

TGFα AUTOCRINE LOOPS IN COLON CANCER

We have identified 2 TGFα autocrine loop phenotypes in colon cancer cells. One phenotype is the classical extracellular autocrine loop as evidenced by the ability of TGFα and EGF_r antibodies to inhibit cell proliferation in growth curves and block DNA synthesis in mitogenesis assays (Ziober et al., in press b). This type of autocrine loop is restricted to the Group III cell lines described above which, like normal cells, are dependent upon exogenous growth factors (included EGF itself) for optimal growth and maximal DNA synthesis.

In contrast to the Group III cell lines, growth factor independent Group I cells show an intracrine TGFα loop similar to those described above for IL-3 and PDGF. Group I cell lines express similar amounts of TGFα as many of the Group III cell lines. However, proliferation and DNA synthesis are not stimulated by exogenous EGF or TGFα in these cell lines. Moreover, TGFα and EGF_r antibodies do not inhibit proliferation or DNA synthesis in this growth regulatory phenotype.

The growth regulatory phenotype of Group I cells suggested an intracrine TGFα phenotype, but it had not been shown that these cells were actually dependent upon TGFα for proliferation and DNA synthesis. In order to prove direct involvement of

TGFα in growth regulation we stably transfected these cells with a vector containing an anti-sense cDNA to the full length TGFα cDNA. Expression of anti-sense TGFα in RNA in these cells led to a 4-5 fold reduction of TGFα sense mRNA and protein expression. In addition, stably transfected cells now required an exogenous source of TGFα or EGF for proliferation and DNA synthesis thus indicating that an intracrine TGFα loop had been disengaged in these cells (Ziober et al., in press a).

The disruption of the intracrine TGFα loop thus led to the restoration of a growth factor dependent phenotype in Group I cell lines similar to that constitutively expressed in Group III cell lines. The restoration of response to environmental growth controls in this cell type suggested that disruption of the TGFα loop might also lead to a reduction of the tumorigenic properties of these cells. Indeed, the anchorage dependent growth of TGFα anti-sense transfected cells was reduced relative to parental cells even in the presence of supplemental EGF or TGFα. Comparison of the TGFα anti-sense transfected cells with parental cells indicated a reduction in the numbers of xenografts generated in athymic nude mice in cells with a repressed TGFα autocrine loop. Moreover when xenografts from TGFα anti-sense transfected cells did form, their growth to a 1 gm tumor was delayed from 2 weeks by parental cells to 4 weeks by transfected cells. Analysis of the tumors which did form indicated that anti-sense mRNA was still expressed, but endogenous levels of TGFα expression had been increased such that net production of endogenous TGFα was higher than that of parental cells despite the presence of anti-sense mRNA. This high level of endogenous TGFα expression was retained when cell lines were generated from these xenografts.

Results with the repression of TGFα in Group I cells suggested that overexpression of the protein in cells with weak TGFα autocrine loops should generate a strong autocrine loop and progression of the recipient cells' tumorigenic properties. We tested this hypothesis in a Group III cell line designated GEO which showed low levels of TGFα expression and weak TGFα autocrine loop as evidenced by the requirement of exogenous EGF for optimal proliferation and the ability of TGFα and EGF_r neutralizing antibodies to inhibit growth of the cells. TGFα sense mRNA was overexpressed in GEO cells to a level at least 25 fold higher than in parental cells. This level of expression abrogated the requirement for exogenous EGF for cell proliferation and rendered the transfected cells growth factor independent with respect to mitogenesis (Ziober et al, in press b). These alterations in growth regulatory phenotype did not lead to any change in doubling time of the transfected cells relative to parental cells. However, transfected cells were 4-10 fold more tumorigenic than parental cells indicating that the formation of a constitutive TGFα autocrine loop leads to progression of the tumorigenic properties of colon cancer cells (Ziober et al., in press b).

These results indicated that TGFα is a viable growth regulatory target for therapeutic intervention in colon cancer, but greater understanding of the factors controlling TGFα expression would be necessary for any eventual application. The results also indicated that therapeutic approaches involving extracellular intervention at the receptor or ligand level are not likely to be effective against progressed growth regulatory phenotypes. Consequently, we began to explore the transcriptional control of TGFα as a possible avenue toward developing a strategy to control TGFα by intracellular agents.

TRANSCRIPTIONAL CONTROL OF TGFα EXPRESSION

Analysis of TGFα transcription was initiated by cloning a 2.8 Kb region 5' of the ATG start site of the human TGFα gene. This region was sequenced and analyzed for consensus sequences corresponding to cis-elements for known transcription factors. As reported by others (Jakobovits et al., 1989; Blasband et al., 1990) the TGFα promoter lacks a TATA box and is highly GC rich. As such, Sp1 sites are common in both the human and rat TGFα promoters and have been shown to be necessary for optimal transcriptional activity (Shin et al., 1992). However, Sp1 represents a cis-element common to many genes. Therefore, while agents which disrupt Sp1 response might lead to reduction of TGFα expression, a large number of other genes would also likely be affected.

It is not likely that there are cis-elements controlling the TGFα gene which are absolutely specific, however, greater specificity for the TGFα gene and the genes affected downstream of TGFα-EGF_r interaction would be desirable. To this end we have concentrated our analyses of TGFα expression on the human autoregulatory response of TGFα to itself with enhanced transcription (Coffey et al., 1987). As indicated above, the rationale for this target is that both TGFα expression and the expression of genes affected by TGFα itself could be down-regulated by the same agent. Moreover, the results with TGFα anti-sense and sense expression described in the previous section demonstrate that it is not necessary to completely eliminate TGFα activity to achieve significant reduction in tumorigenic properties.

AUTOREGULATION OF TGFα TRANSCRIPTION IN COLON CANCER CELLS

TGFα promoter-CAT reporter gene constructs were generated for variously sized deletion fragments of the 2.8 Kb 5' region of the TGFα promoter. Fragments generated included 2.8 Kb, 1.5 Kb, 1.1 Kb and 0.34 Kb 5' of the TGFα ATG start site. Although all of these fragments showed equal activity in breast cancer cells (Saeki et al., 1992), the 1.5 Kb fragment showed enhancer activity and the 2.8 and 1.1 Kb fragments showed repressor activity in colon carcinoma cell lines relative to the basal activity of the 0.34 Kb fragment (Table 2). Reduction of the 0.34 Kb fragment led to progressive loss of promoter activity.

TABLE 2: RELATIVE TGFα PROMOTER ACTIVITY IN CELL LINES

	FRAGMENT (Kb)			
CELL LINE	0.34	1.1	1.5	2.8
HCT 116	1	0.8	1.3	0.7
FET	1	0.7	1.5	0.5
GEO*	1	1	1	1

*GEO TGFα promoter activity was approximately 10% that of HCT 116 and FET

We hypothesized that TGFα transcription would be dependent in part upon the strong TGFα autocrine loop expressed by Group I cells since others had shown that the exposure of cells to exogenous EGF or TGFα could lead to enhanced TGFα transcription (Coffey et al., 1987). In order to test this hypothesis we initially compared transcription in a Group I cell line designated HCT 116 with the corresponding HCT116 cells which had been transfected with TGFα to disengage the TGFα autocrine loop. As expected the TGFα anti-sense transfected cells, designated HCT116 U, showed reduced TGFα transcription relative to HCT116 cells using any of the 4 TGFα promoter-CAT reporter fragments described above. Levels of TGFα transcription by the 0.34 Kb fragment in HCT116 U cells were approximately 33% those of HCT116 cells. TGFα transcription in HCT116 U cells could be partially, but not completely restored by exogenous TGFα treatment. Similar results were obtained with the longer TGFα promoter-reporter fragments.

As indicated above cell lines isolated from the xenografts which developed from TGFα anti-sense transfected cells were isolated which expressed larger amounts of endogenous TGFα then the parental cells. HCT116 UX was the designation of the cell line isolated from xenografts formed by HCT116 U cells and TGFα mRNA expression in this cell line was approximately 1.3 fold that of HCT116 cells. The TGFα levels in HCT116 UX cells suggested that increased TGFα transcription might be reflected by high TGFα expression in these cells. Transcription rates in HCT116 UX cells as well as the other cell lines isolated from xenografts of TGFα anti-sense transfected cells were slightly higher than in HCT116 cells and several fold higher than in the TGFα anti-sense transfected cells prior to xenograft formation. Thus transcription rates of TGFα in colon cancer cells appear to be highly dependent upon the formation of a strong autocrine loop while increased transcription appears to contribute to the reformation of a strong autocrine loop in xenografts from TGFα anti-sense transfected cells (Table 3).

TABLE 3: TGFα TRANSCRIPTION LEVELS IN HCT 116 CELLS AND TGFα ANTI-SENSE CLONES

CELL	RELATIVE TRANSCRIPTION (%)
HCT 116	100
HCT 116 U*	25
HCT 116 UX*	130
GEO	10
GEO-TGFα**	55

*HCT 116 U transfected with TGFα anti-sense.
HCT 116 UX cell line isolated from HCT 116 U xenograft.
**Cells transfected with TGFα sense.

TRANSCRIPTIONAL STIMULATION OF TGFα BY EXOGENOUS EGF OR TGFα

The results described above indicated that TGFα transcription levels in Group I colon carcinoma cell lines were dependent upon a strong TGFα autocrine loop and were not influenced by exogenous EGF or TGFα in the tissue culture medium. Therefore, it was of interest to determine whether transcription of TGFα in Group III cells with weak TGFα loops was dependent upon exogenous growth factors. TGFα promoter-reporter assays were performed with the 0.34 Kb and 1.5 Kb constructs in the presence and absence of EGF in the tissue culture medium of a Group III cell line designated FET. Steady state transcription was approximately 3 fold higher for both constructs in the FET cells maintained in EGF. FET cells were stably transfected with the 0.34 bp promoter-reporter construct to determine the kinetics of EGF induction of TGFα transcription. Increased transcription was observed after 2 hrs and stabilized by 4 hrs. TGFα mRNA levels paralleled increased transcription. These results indicated that exogenous growth factors control TGFα transcription in these relatively unaggressive unprogressed colon carcinoma cell types.

LOCALIZATION OF THE AUTOSTIMULATORY DNA SEQUENCE

Since differences in transcription between HCT116 and HCT116 U cells were parallel in all 4 of the TGFα promoter-reporter constructs the specific DNA sequence responsible for TGFα autostimulation of transcription appeared to be in the 340 bp fragment. Therefore additional deletions were performed on the 340 bp promoter fragment. It was found that a 201 bp fragment gave equal transcription in HCT116 and HCT116 U cells while a 247bp gave higher transcription in HCT116 cells than HCT116 U cells. This indicated that the DNA sequence responsible for TGFα autostimulation was between 201 and 247bp 5' of the ATG start site. Overlapping 15 mer oligonucleotides were synthesized to span the region between 201 and 247 and then characterized in heterologous promoter-reporter assays, DNA gel shift assays and Southwestern blots.

The overlapping oligonucleotides were placed upstream of a thymidine kinase (TK) promoter-CAT reporter construct to determine whether the putative TGFα autostimulatory element would impart the same responsiveness to changes in the TGFα autocrine loop as the TGFα promoter-reporter constructs containing the response element. One of the 15 mer oligonucleotides was able to stimulate transcription of the TK promoter relative to the TK promoter without the oligonucleotide in HCT116 cells while no stimulation was observed in HCT116 U cells. Exogenous TGFα treatment led to the stimulation of this GC rich oligonucleotide-TK promoter construct but not the TK promoter alone in HCT116 U cells. As expected HCT116 UX cells showed higher TK promoter transcription with constructs containing the oligonucleotide. The same oligonucleotide was able to mediate increased transcription of the TK promoter in response to EGF treatment of FET cells, thus indicating that this DNA sequence is reponsible for autoregulation of transcription by TGFα as well as by exogenous activation of the EGF_r.

Cis-elements (the DNA sequences controlling transcriptional response) interact with proteins called trans-activating factors to enhance or repress transcription. One might expect that nuclear extracts from cells with greater levels of transcription would show higher levels of proteins with specific binding to a cis-element of interest if transcription is enhanced. Specific DNA binding proteins can be recognized by gel shift analysis in which labeled DNA oligonucleotides are mixed with proteins from nuclear extracts followed by gel electrophoresis. Free DNA is allowed to run to the end of the gel. The larger the amount of specific DNA binding protein available in nuclear extracts the more labeled probe that is retained by the proteins in the gel. Gel shift analysis with the 15 mer TGFα autostimulatory oligonucleotide element showed that more DNA binding protein was present in nuclear extracts from HCT116 cells and HCT116 UX cells than in HCT116 U cells. Moreover Sp1 consensus sequences were not able to compete for binding of proteins to the oligonucleotide indicating that the GC rich TGFα responsive DNA element does not bind to Sp1 protein which, as indicated above, is a known stimulator of TGFα transcription.

Southwestern analysis is used to identify the molecular weight of proteins which bind to DNA elements. In this technique proteins from nuclear extracts are subjected to gel electrophoresis, transferred to nitrocellulose and then labeled DNA oligonucleotide probe is added in much the same manner as Western analysis by antibodies to identify those proteins to which the DNA sequence of interest binds. The oligonucleotides spanning the region between 201 and 247bp showed binding to 3 proteins in Southwestern analysis. The oligonucleotide representing the putative TGFα response element bound to proteins of 105, 47 and 25 Kd while the other oligonucleotides did not bind to the 47 Kd protein. These results suggest that the 47 Kd protien plays a critical role in the control of TGFα transcription as a potential trans-acting factor. Efforts are currently underway to isolate and characterize this protein.

INTERACTION OF THE TGFα AUTOCRINE LOOP WITH A GASTRIN AUTOCRINE LOOP

Previous studies from our laboratory showed that gastrin receptor antagonists or anti-gastrin antibodies were capable of inhibiting colon carcinoma cell growth (Hoosein et al., 1988, 1990). Others had shown that gastrin transcription is enhanced by EGF. Therefore, we hypothesized that autocrine gastrin expression might be under autocrine TGFα control in HCT116 colon carcinoma cells. If this hypothesis were correct it would be expected that HCT116 U cells would show reduced gastrin expression relative to HCT116 cells because of the repression of TGFα in the anti-sense transfected cells. A human specific gastrin RNAse protection assay showed that steady state gastrin mRNA levels in HCT116 U cells were ~ 33% those of HCT116 cells. Like TGFα, gastrin expression in HCT116 UX cells was 1.3 fold higher than HCT116 cells. These results indicated that disengagement of the TGFα autocrine loop or repression of TGFα transcription would not only lead to inhibition of cancer growth mediated by TGFα, but also gastrin.

TGFβ ACTS AS A TUMOR SUPPRESSIVE FACTOR

TGFβ is a hormonal like polypeptide which is inhibitory to many cell types including some cancer cells. We found that Group III colon cancer cells could be inhibited by exogenous TGFβ treatment, but Group I cells are completely refractory to the polypeptide (Hoosein et al., 1987, 1989). It has been hypothesized that progression of malignancy may involve the loss of autocrine anti-proliferative effects by TGFβ (Wakefield and Sporn, 1990). Several investigators had treated TGFβ responsive cancer cells with TGFβ neutralizing antibodies and shown increased proliferation or DNA synthesis resulted indicating that these cells did produce autoinhibitory TGFβ (Arteaga et al., 1990; Hafez et al., 1990). However, there were no studies indicating whether malignancy was affected by autocrine TGFβ. Consequently, we explored the tumor suppressive activity of TGFβ using a TGFβ anti-sense approach to constitutively repress $TGF\beta_1$ expression in a Group III colon carcinoma cell line (designated FET) which expresses only the $TGF\beta_1$ gene of the 3 known human TGFβ genes (Wu et al., 1992). FET cells like the other Group III colon carcinoma cell lines are poorly tumorigenic in athymic mice (Brattain et al., 1984). $TGF\beta_1$ anti-sense transfected FET cells showed a level of tumorigenicity similar to that of Group I cells. This was the expected result if $TGF\beta_1$ had a tumor suppressive role in FET cells. Similar results were obtained with a second Group III cell line designated CBS which expressed the $TGF\beta_2$ as well as the $TGF\beta_1$ gene. Presumably, the high degree of homology between $TGF\beta_1$ and B_2 allowed for $TGF\beta_2$ interaction with $TGF\beta_1$

anti-sense mRNA. Thus, the hypothesis that loss of autocrine negative TGFβ activity leads to malignant progression appears to be valid with respect to tumorigenicity in athymic mice.

REPRESSION OF TGFβ LEADS TO GROWTH FACTOR INDEPENDENCE

These results raised the issue of the mechanism of this progression. Given the role of growth factor independence in progression of malignancy we hypothesized that autocrine TGFβ may function in part by maintaining growth factor dependence for DNA synthesis. Previous work had shown that growth factors did not have any effect on the stimulation of DNA synthesis in Group I cells, i.e. nutrient replenishment alone was sufficient to initiate DNA synthesis in quiescent cells. This was in contrast to Group III cells such as FET and CBS which required exogenous growth factors to obtain maximal DNA synthesis after the establishment of quiescience (Mulder and Brattain, 1989). However, Group I and Group III cells showed the same cell cycle time and had the same doubling times in cell culture.

Group III cells transfected by TGFβ anti-sense did not show a change in doubling time, but they were less sensitive to lag time effects when plated at low inocula in tissue culture (Wuu et al., 1992). This suggested that autocinre TGFβ might act by maintaining a quiescent state in untransfected Group III cells unless a sufficient level of stimulatory growth factors were present to overcome the block to DNA synthesis. We tested this hypothesis by comparing the growth factor requirement for release of TGFβ anti-sense transfected FET and CBS cells to control cells. Anti-sense transfected cells were more difficult to render quiescent than untransfected cells. FET and CBS cells show minimal incorporation of ^{3}H thymidine incorporation within 5 days of factor removal and nutrient depletion whereupon less than 10% of cells show ^{3}H thymidine uptake by autoradiography. Minimal incorporation of ^{3}H thymidine is not obtained until 7 days for anti-sense transfected cells. Moreover, maximal DNA synthesis after release from quiescence was obtained by nutrients alone in the anti-sense transfected cells (Table 4). Therefore, autocrine negative TGFβ helps induce quiescence and maintains a growth factor dependent state for release from quiescence.

TABLE 4: EFFECTS OF TGFβ ANTI-SENSE TRANSFECTION

A. ESTABLISHMENT OF QUIESCENCE

CELL	10^6 CPM ^{3}H THYMIDINE/10^6 CELLS ON DAY: 2	4	6
Control	0.20	0.07	0.05
Anti-Sense	0.45	0.20	0.10

B. RELEASE FROM QUIESCENCE

CELL	FOLD INCREASE IN CMP ^{3}H THYMIDINE/10^6 CELLS AFTER ADDITION OF: NUTRIENTS ALONE	NUTRIENTS + GROWTH FACTORS
Control	1.1	7.2
Anti-Sense	10.0	10.2

INTERACTION OF TGFα AND TGFβ AUTOCRINE LOOPS

The growth factor independence resulting from repression of autocrine TGFβ was also consistent with malignant progression. Given the importance of TGFα autocrine loops in the malignant progression of colon carcinoma cells we hypothesized that the growth factor independence of the TGFβ anti-sense transfected cells might be due to increased TGFα autocrine activity. Consequently, we examined the role of autocrine TGFα in the establishment of quiescence and mediation of DNA synthesis in the release from quiescence in colon carcinoma cells.

As quiescence in untransfected CBS and FET cells is established, TGFα and EGF_r mRNA expression is reduced (Mulder et al., 1990a). TGFβ anti-sense transfected FET and CBS cells also showed increasing TGFα and EGF_r as quiescence was established. This suggested that growth factor independence might be due to the generation of increased TGFα activity during the establishment of quiescence (Mulder, 1991). In order to test this hypothesis TGFβ anti-sense transfected cells were treated with EGF_r antibody during release from quiescence.

The rationale was that if the nutrient replenishment response of anti-sense transfected cells was due was mediated by the generation of a stronger TGFα autocrine loop by TGFβ repression, the EGF_r antibody should act to block the TGFα autocrine activity and thus block the nutrient replenishment mediated DNA synthesis of the transfected cells. EGF_r antibody treatment led to a reduction of more than 50% of nutrient replenishment mediated DNA synthesis in TGFβ anti-sense transfected cells but had little if any effect on untransfected cell nutrient replenishment response. These results indicated that one function of autocrine TGFβ activity is control of TGFα autocrine loop expression during the establishment of quiescence. Moreover, the control of the TGFα autocrine loop by autocrine TGFβ appears to be a key function for insuring that DNA synthesis is mediated by environmental growth factors after cells have entered a quiescent state.

PROSPECTS FOR ANTI-CANCER AGENTS WITH TRANSCRIPTIONAL THERAPEUTIC TARGETS

The process of discovery and development of anti-cancer drugs carries with it many practical considerations. In general, cost must be balanced against the prospects for success which for cancer would be a drug which is broadly applicable to a number of cancers with increased survival. Model systems for predicting the clinical performance of cancer drugs provide little comfort for the investigator with respect to the certainty of success or failure. Pragmatically this has the effect of raising the value of cost considerations in determining whether to pursue a potential individual drug or class of drugs. Any discussion of cost is particularly germaine to biological and molecular biological agents which are extremely expensive to develop and supply. The expense of developing biological agents would also reflect high expenses related to treatment itself and may affect third party cost considerations. Thus, the development of any biological agent will ultimately require a high degree of confidence for success according to the cancer drug standards described above.

Over the years the exploitation of most biological targets has involved strategies involving the supply of a protein in the case of a target involving a missing activity or for the provision of an inhibitory activity. The clinical trials of interferon are an example of this approach. The most common biological approaches have involved the use of antibodies either to putative cancer specific antigens or to biologically active targets such as the EGF_r. Both of these extremely expensive approaches have been disappointing for a variety of reasons which will not be reviewed here.

It is not the intent of this discussion to say that these types of approaches will never be viable, but rather to point out that there is both a scientific and a practical need for small molecular weight agents which can either agonize or antagonize biological activities relevant to cancer in much the same way pharmacological agents have been developed for diseases other than cancer. Much of this development for other diseases was based on identifying agonists and anatagonists for specific molecular targets.

This presentation has described the importance of TGFα to colon cancer progression and growth, both _in vitro_ and _in vivo_ as xenografts. Moreover, the importance of TGFα transcriptional autoregulation has been described such that we are beginning to understand that this autocrine factor has complex interactions with other autocrine growth factor systems. Thus, disruption of TGFα transcriptional autoregulation leads to significant reduction of malignant properties without killing the target cell.

Interactions between trans-acting factors and cis-elements provide a potentially excellent system for molecular targets which could be modeled either _in vitro_ or in more complex systems involving cells and animals. For example, screens for small molecular weight molecules such as natural products which interfere with TGFα transcription could involve an _in vitro_ transcription assay of the TGFα response element attached to the TK promoter-CAT reporter construct, a cell line such as HCT116 transfected with the construct or a xenograft with the HCT116 cell line transfected with the construct. Alternatively, the interactions of cis and trans elements are extremely sensitive to structural changes. For example the Wilm's tumor product can be either a repressor or enhancer of gene function depending upon whether 3 or 4 of its zinc fingers interact with its cis-element. Such interactions should be amenable to molecular modeling approaches.

ACKNOWLEDGEMENTS

Supported by NIH Grants: CA4967, CA50457, CA38173, CA34432, CA54807 (MGB) and CA51452 and 54816 (KMM). The authors thank Ann Chlebowski for preparation of the manuscript.

LITERATURE REFERENCES

Ankesaria P, Teixido J, Laiho M, Pierce JH, Greenberger JS, Massague J (1990) Cell-cell adhesion mediated by binding of membrane-anchored transforming growth factor-α to epidermal growth factor receptors promotes cell proliferation. Proc. Natl. Acad. Sci. USA 87:3289-3293

Arteaga CL, Coffey RJ, Dugger TC, McCutchen CM, Moses HL, Lyons RM (1990) Growth stimulation of human breast cancer with anti TGFβ antibodies: evidence for negative autocrine regulation by TGFβ. Cell Growth and Differentiation 1:367-374

Bates SE, Valverius EM, Ennis BW, Bronzert DA, Sheridan JP, Stampfer MR, Mendelsohn J, Lippman ME, Dickson RB (1990) Expression of the transforming growth factor-alpha/epidermal growth factor receptor pathway in normal human breast epithelial cells. Endocrinology 126:596-607

Blasband AJ, Rogers KT, Chen X, Azizkhan JC, Lee DC (1990) Characterization of the rat transforming growth factor-α gene and identification of promoter sequences. Mol. Cell. Biol. 10:2111-2121

Boyd D, Levine AE, Brattain DE, McKnight MK, Brattain MG (1988) A comparison of growth requirements of two human intratumoral colon carcinoma cell lines in monolayer and soft agarose. Cancer Res. 48:2469-2474

Brachmann R, Linquist PB, Nagashima M, Kohr W, Lipari T, Napier M, Derynck R (1989) Transmembrane TGF-α precursors activate EGF/TGF-α receptors. Cell 59:691-700

Brattain MG, Levine AE, Chakrabarty S, Yeoman LC, Willson JKV, Long BH (1984) Heterogeneity of human colon carcinoma. Cancer Metastasis Rev 3:177-191

Browder TM, Dunbar C, Nienihuis A (1989) Private and public autocrine loops in neoplastic cells. Cancer Cells 1:9-17

Chantret I, Barbat E, Dussaulx E, Brattain MG, Zweibaum A (1988) Epithelial polarity, villin expression, and enterocyte differentiation of cultured human colon carcinoma cells: a survey of twenty cell lines. Cancer Res. 48:1936-1942

Coffey RJ, Derynck R, Wilcox JN, Bringman TS, Goustin AS, Moses HL, Pittelkow MR (1987) Production and auto-induction of transforming growth factor-α in human keratinocytes. Nature 328:817-820

Derynck R, Goeddel DVM, Ullrich A, Gutterman JU, Williams RD, Bringman TS, Berger WJ (1987) Synthesis of mRNAs for transforming growth factor-α and β and the epidermal growth factor receptor by human tumors. Cancer Res. 47:707-712

Gill G, Kawamoto T, Cochet C, Le AD, Sato JD, Musui H, McLeod C, Mendelsohn J (1984) Monoclonal anti-epideral growth factor receptor antibodies which are inhibitors of epidermal growth factor binding and antagonists of epidermal growth factor-stimulated tyrosine protein kinase activity. J. Biol. Chem. 259:7755-7760

Hafez MM, Infante D, Winawer S, Friedman E (1990) Transforming growth factor-β_1 acts as an autocrine negative growth regulator in colon enterocytic differentiation but not in goblet cell maturation. Cell Growth and Differentiation 1:617-626

Hoosein NM, Brattain DE, Matthews MK, Levine AE, Brattain MG (1987) Characterization of the inhibitory effects of transforming growth factor-β on a human colon carcinoma cell line. Cancer Res. 47:2950-2954

Hoosein NM, Kiener PA, Curry RC, Rovati LC, McGilbra DK, Brattain MG (1988) Antiproliferative effects of gastrin receptor antagonists and antibodies to gastrin on human colon carcinoma cell lines. Cancer Res. 48:7179-7183

Hoosein NM, McKnight MK, Levine AE, Mulder KM, Childress KE, Brattain DE, Brattain MG (1989) Differential sensitivity of subclasses of human colon carcinoma cell lines to the growth inhibitory effects of transforming growth factor-β_1. Experimental Cell Research 181:442-453

Hoosein NM, Kiener PA, Curry RC, Brattain MG (1990) Evidence for autocrine growth stimulation of cultured colon tumor cells by a gastrin/cholecystokinin-like peptide. Exp. Cell Res. 186:15-21

Jakobovits EG, Schlokat U, Vannice J, Derynck R, Levinson AD (1988) The human TGFα promoter directs transcription initiation from a single site in the absence of a TATA sequence. Mol. Cell. Biol. 8:5549-5554

Massague J (1990) Transforming growth factor α: A model for membrane anchored growth factors. J. Biol. Chem. 265:21393-21396

Merchant JL, Denediuk B, Brand SJ (1991) A GC-rich element confers epidermal growth factor responsiveness to transcription from the gastrin promoter. Mol. Cell Biol. 11:2686-2696

Mulder KM (1991) Differential regulation of c-myc and transforming growth factor-α messenger RNA expression in poorly-differentiated and well-differentiated colon carcinoma cells during the establishment of a quiescent state. Cancer Res. 51:2256-2262

Mulder KM, Brattain MG (1989) The cell and molecular biology of colon cancer. L.E. Augenlicht (ed). CRC Press, Boca Raton, FL, pp. 45-67

Mulder KM, Brattain MG (1989) Effects of growth stimulatory factors on mitogenicity and c-myc expression in poorly-differentiated and well-differentiated human colon carcinoma cells. Mol. Endocrinol. 3:1215-1222

Mulder KM, Humphrey LE, Choi HG, Childress-Fields KE, Brattain MG (1990) Evidence for c-myc in the signalling pathways for TGF-β in well-differentiated human colon carcinoma cells. J. Cell. Physiol. 145:501-507

Mulder KM, Zhong Q, Choi HG, Humphrey LE, Brattain MG (1990) Inhibitory effects of transforming growth factor-β_1 on mitogenic response, transforming growth factor-α, and c-myc in quiescent, well-differentiated human colon carcinoma cells. J. Cell. Physiol. 145:501-507

Saeki T, Cristiano A, Lynch MJ, Brattain M, Kim N, Normanno N, Kenney N, Ciardello F, Saloman DS (1991) Regulation by estrogen through the 5' flanking region of the TGFα gene. Mol. Endo. 5:1955-1963

Shin TH, Paterson AJ, Grant JH, Meluch AA, Kudlow JE (1992) 5-Azalytidine treatment of HA-A melanoma cells induces Sp1 activity and concomitant TGFα expression. Mol. Cell. Biol. 12:3998-4006

Sporn MB, Todaro GJ (1980) Autocrine secretion and malignant transformation of cells. N. Engl. J. Med. 303:878-880

Wakefield LM, Sporn MB (1990) In tumor suppressor genes. G. Klein (ed).

Wan CW, McKnight MK, Brattain DE, Brattain MG, Yeoman LC (1988) Different epidermal growth factor responses and receptor levels in human colon carcinoma cell lines. Cancer Letters 43:139-143

Wong ST, Winchell LF, McClune BK, Earp HS, Texido J, Massague J, Herman B, Lee DC (1989) The TGF-α precursor expressed on the cell surface binds to the EGF receptor on adjacent cells, leading to signal transduction. Cell 56:495-506

Wu SP, Theodorescu D, Kerbel R, Willson JKV, Mulder KM, Humphrey LE, Brattain, MG (1992) TGF-β_1 is an autocrine negative growth regulator of human colon carcinoma FET cells <u>in vivo</u> as revealed by transfection of an anti-sense expression vector. J. Cell. Biol. 116:186-197

Ziober BL, Willson JKV, Humphrey LE, Childress-Fields KE, Brattain MG (1992) Evidence for an intracellular TGF-α autocrine loop in HCT 116 colon carcinoma cells. J. Cell Biol. in press.

Ziober BL, Willson, JKV, Humphrey LE, Fields KC, Brattain MG (1992) Autocrine TGFα is associated with progression of transformed properties in human colon carcinoma cells. J. Biol. Chem. in press.

INTERFERON REGULATION OF DIFFERENTIATION AND MECHANISMS

Rossi G.B.[1], Romeo G.[1,2], Battistini A.[1], Affabris E.[1,3], Coccia E.M.[1], Fiorucci G.[1,2]
[1]Laboratory of Virology, Istituto Superiore di Sanità; [2]Istituto Tecnologie Biomediche, CNR; [3]III Università degli Studi di Roma.

ABSTRACT

The clear involvement of the interferon (IFN) system in the regulation of differentiation justifies the advanced IFN clinical trials in "differentiation therapy" of cancer. By employing this strategy, it may be possible to specifically reprogram the phenotype of a tumor cell by inducing its terminal cell differentiation and, thus, the loss of its tumorigenic potential. The effects of IFNs on erythroid differentiation and the related mechanisms are reported. Administration of highly purified preparations of murine IFN-alpha or -beta to Friend leukemia cells, induced to differentiate by dimethyl-sulfoxide, leads to a 100% increase of benzidine-positive cells. Both species of IFN induce a substantial increase in heme, hemoglobin and transferrin receptor levels. The results obtained suggest that in erythroid cells the intracellular heme level may represent a key regulatory factor in the hemoglobin synthesis pathway. It is postulated that IFN induces the enhancing effect on differentiation via a marked increase of heme synthesis and number of transferrin receptors which in turn leads to an enhancement of globin chain synthesis.

INTRODUCTION AND BACKGROUND

Interferon(s) (IFNs) are members of a network of substances (now called cytokines) that are all able to operate as regulatory molecules in the homeostatic control of cellular functions. Cytokines may be produced constitutively at low levels and exert multiple effects on virtually all cells. They are active participants in host defenses against viral or parasitic infections and tumors. It is now well accepted that IFNs affect normal cell division and many specialized cellular functions.

Research into IFN and cytokines is expanding markedly, and there are increasing therapeutic applications for cytokines. In fact, as reviewed below, the involvement of the IFN system in the regulation of differentiation has also led to significant advances in planning IFN clinical trials and cancer therapy. This alternative approach involves the use of agents which are not directly cytotoxic but modify tumor cell growth by inducing terminal cell differentiation, i.e., loss of proliferative capacity without a concomitant loss of cell viability. This strategy lies in the belief that neoplasia originates from the inhibition of cell differentiation. If this inhibition is overcome, a reprogrammed phenotype of a tumor cell, with possible loss of tumorigenic potential, would ensue. The ability to employ the above methodologies for cancer therapy requires the development of appropriate model

NATO ASI Series, Vol. H 75
Cancer Therapy
Edited by N. D'Alessandro, E. Mihich, L. Rausa, H. Tapiero, and T. R. Tritton

systems and the identification of both single and multiple agents capable of inducing terminal differentiation of tumor cells. The present review will discuss how IFNs may function as regulatory molecules and how they may alter cell differentiation by either inhibiting or inducing it, depending on the target cells.

Although extensively studied, the relationships between the multiple effects of IFNs and their pathways of action remain to be defined. For example, following the interaction of IFN with its appropriate cell-surface receptor, the nature of the signal transduction mechanism mediating the pleiotropic effects of IFN is not clear. Recent studies have shed some light on these early events regarding both primary signal-transducing agents (i.e., protein kinases, phosphatidylinositol or sphyngomyelin turnover, etc.) and the activation of specific transcriptional changes in target cells (i.e., activation of latent transcription binding factors able to bind to the IFN-stimulated-response-element, ISRE, an enhancer element inducible by IFNs for transcriptional activation). It is tempting to postulate that some regulatory functions attributed to IFN are mediated through IFN-induced enzymes, generally referred to as the IFN system, such as 2'-5' oligoadenylate (2-5A) synthetase, 2-5A-activated RNase and double-stranded (ds) RNA-activated protein kinase.

IFN IN DIFFERENTIATING AND DIFFERENTIATED CELL SYSTEMS

Cell differentiation depends upon a program of ordered gene expression resulting in the production of specific proteins, usually associated with a specific, terminally differentiated phenotype. IFN genes in undifferentiated stem cells of teratocarcinoma (embryonal carcinoma cells) are refractory to virus induction. In addition, these cells and those of early stage embryos are not sensitive to IFN action. In this and other systems (Harada et al., 1990), genes responsible for the induction and action of IFNs are in a repressed state and they become functional only after cell differentiation. Also IRF-genes (coding for protein factors that specifically bind the IFN-alpha and -beta gene promoters, as well as the ISRE of IFN-stimulated genes) are developmentally controlled (Harada et al., 1990). The effects elicited by IFN on the complex biologic phenomena related to cell differentiation, i.e., either stimulatory or inhibitory, depend not only on the cell system employed, but also on the type of IFN used (Rossi, 1985; Rossi et al., 1987; De Maeyer and De Maeyer-Guignard, 1989; Romeo et al., 1989). Although the reason for the differential effects of IFN on specific target cells is not known, the diverse responses to IFN may reflect differences in the type of transmembrane signals elicited by different types of IFNs following their binding to the appropriate cell membrane receptor and/or by the various target cells.

Several culture systems of differentiating or differentiated cells have been studied with respect to the effects of IFNs. In general, these differentiation systems appear to be profoundly affected by exposure to IFN. The effects have been highly specific, in that they are not accompanied by any modification of overall cellular protein synthesis. This observation is important to dispel any residual disbelief about the selectivity of IFN action.

Inhibition of adipose conversion of Balb/c 3T3 and of 3T3-L1 mouse fibroblasts by type I IFN has been reported by Cioè et al., (1980) and Keay and Grossberg (1980). Inhibition of the exase monophosphate shunt activity has been proposed as the mechanism of blocking differentiation of preadipose cell lines (Saneto and Johnson, 1982).

With respect to melanogenesis, Fisher et al. have reported that IFN-beta inhibits differentiation in murine B-16 melanoma cells, whereas similar IFN preparations induce melanogenesis in specific human melanoma cell lines. In human melanoma cell cultures, the combination of IFN-beta and mezerein results in an induction of terminal differentiation, with a concomitant loss of proliferative capacity and an increase in the synthesis of melanin (Ahmed et al., 1989). Induction of terminal differentiation by the combination of IFN-beta and mezerein in human melanoma cells is associated with the altered expression of several cellular genes and changes in the expression of specific cell surface antigens (Ahmed et al., 1989; Graham et al., 1991). Now it appears (Jiang et al., 1992) that this chemical induction of differentiation in human melanoma cell is a two-stage process consisting of early gene expression changes and a subsequent autocrine feed-back mechanism. In relation to the preceding findings, Levitt et al. (1990) reported that IFN-beta selectively induces squamous differentiation and growth suppression in squamous cell differentiation competent, lung tumor cell lines.

Treatment of human epidermal keratinocytes with IFN-gamma results in irreversible growth arrest and induction of a squamous differentiated phenotype. IFN-gamma-induced squamous differentiation was characterized by an increase in the expression of squamous cell-specific genes. A combination of retinoic acid and hu rec IFN-gamma led to a pronounced synergistic amplification of growth inhibition in cultured breast cancer cells. Combined systemic therapy with retinoic acid and IFN-alpha-2a is highly effective in patients with advanced squamous cell carcinoma of the skin (Lippman et al., 1992). We reported before (Improta et al., 1988) that IFN-gamma causes a reversible arrest of proliferation of neuronal cells PC-12 thus facilitating NGF-induced normal differentiation of these cells.

The ability of IFN to modulate myogenesis in human myoblast cultures has been observed (Andre et al., 1988; Fisher et al., 1983). Andre et al. (1988) reported that rat IFN enhances the expression of acetylcholine receptors in rat myotubes in culture. Hu-rIFN-alphaA was found by Fisher et al. to induce an acceleration of myotube formation and creatine kinase isoenzyme transition in normal human myoblast cultures derived from mature skeletal muscle. In chickens, however, IFN treatment inhibits the differentiation of embryo myogenic cells (Tomita and Hasegawa, 1984). The discrepancy of these observations on IFN action on myotube formation may be more apparent than real as the culture systems were derived from mature and embryonic tissues, respectively.

More recently the involvement of the IFN system has also been studied (i.e., the 2-5A synthetase and dsRNA-dependent protein kinase pathways) in the appearance of a variety of proteins related to the formation of differentiated muscle fibers, that appear to be muscle-specific proteins or proteins responsible for cell growth (see paragraph on the IFN-inducible enzymes) (Birnbaum et al., 1990).

Hematopoiesis

Hematopoiesis research is certainly the area in which most data on the effects on differentiation have accumulated, and accordingly requires extensive discussion.

Available experimental data support a possible in vivo regulatory action of IFN on the hematopoietic system, although its effects differ from system to system. In vitro treatment with IFN of normal hematopoietic progenitors induces inhibitory effects on CFU-GM, -G, -M, BFU-E and CFU-E development (Rossi et al., 1987; De Maeyer and De Maeyer-Guignard, 1989). An inhibition of stem cell development is frequently observed at the relatively low IFN concentration of a few units per ml. This finding suggests that IFN may exert this effect under physiological conditions in the microenvironment of the spleen or bone marrow, in which low amounts of IFN are frequently present. The antiproliferative effects suggest a role for IFNs in the physiological control of progenitor cell proliferation.

Fibroblast IFN preparations have been suggested as having two distinct actions on bone marrow cells supplemented with macrophage-derived colony-stimulating factor: i) inhibition of growth of early precursor cells (detected in semi-solid cultures), and ii) stimulation of growth and differentiation events at later stages of monocytic differentiation (detected only in liquid cultures) (Yamamoto-Yamaguchi et al., 1983). Perussia et al. 1983 reported that a component of phytohemagglutinin-stimulated, conditioned medium, identified as IFN-gamma, induced terminal monocytic differentiation of human immature myeloid cells from normal bone marrow. A preferential stimulatory effect of IFN-gamma on monocytic differentiation was also confirmed by the other authors (Maciejewski et al., 1990).

In keeping with the inhibitory effect on the growth of hematopoietic progenitors, erythropenia and/or leukopenia have been frequently observed in clinical trials carried out with HuIFN-alpha. The involvement of HuIFN-gamma or IFN-alpha in the development of some cases of aplastic anemia has been also suggested (De Maeyer and De Maeyer-Guignard, 1988).

Several lines of evidence obtained in vitro indicate that IFN modulates the differentiation potential of leukemic cells. In the myeloid lineage, HuIFN-alpha (both native and recombinant) and IFN-beta induce cells of the histiocytic lymphoma line U937 to move along the monocytic pathway of differentiation (Hattori et al., 1983). Testa et al. (1988) observed that HuIFN-alpha and -beta may participate in the inhibition of cell proliferation occurring during cellular differentiation of U937 cells, but only IFN-gamma may be involved in the induction of the expression of specific monocytic markers involved in cellular immunoregulation. More recently, IFN-gamma was reported to be able to abrogate the differentiation block in v-myc-expressing U-937 (Oberg et al., 1991).

An interesting finding is that combination of IFN with differentiation-inducing agents, including 12-o-tetradecanoyl -phorbol-13 acetate (TPA), retinoic acid, mezerein and dimethyl sulfoxide, can potentiate growth suppression and induction of differentiation. This effect is observed in cells which are sensitive, innately resistant, or selected for resistance to growth suppression and induction of differentiation by either agent used alone. Thus, IFN has the ability to potentiate the growth-suppression activity and the differentiation-induction capacity of diverse differentiation-inducing agents. This suggests that, in certain situations, IFN plus a differentiation inducer may prove more beneficial in inhibiting tumor growth than either agent alone. For example, the human promyelocytic leukemia cell line HL-60 cannot be induced to differentiate by IFN-alpha or -beta treatment alone. However, induction of differentiation in vitro by chemical inducers (TPA, retinoic acid, etc.) can in fact be stimulated by IFN-alpha, -beta or -gamma (Tomida et al., 1982; Weinberg et al., 1986). Similarly, mouse myeloid leukemic M1 cells cannot be induced to differentiate into macrophages and granulocytes by IFNs. Nonetheless, IFNs "stimulate" differentiation events, induced by a number of compounds (D-factor, lipopolysaccharide, poly I:C) (Tomida et al., 1980).

Mouse M_1 cells exposed to IL-6 stop growing and differentiate into macrophages in 3-4 days, during which IFN-dependent genes are strongly induced, IL-6 acting by synergism with low amounts of autocrine IFN (Cohen et al., 1991). Differences between IFN-alpha or -gamma effects on the induction of differentiation in primary cultures of myeloid leukemia cells have been reported (Nakamaki et al., 1990). In particular IFN-alpha enhances granulocytic differentiation and IFN-gamma induces mono-macrophage differentiation of promyelocytic leukemic cells in the presence of retinoic acid (Nakamaki et al., 1990).

In the Daudi cell line, a B-cell lymphoblastoid line transformed by Epstein Barr Virus and highly sensitive to IFN, the cell growth inhibition induced by Hu lymphoblastoid or rec-IFN-$alpha_2$ is accompanied by plasmacytoid differentiation (Exley et al., 1987). Cells from other leukemias of B-cell lineage can also be directed toward differentiation by IFN treatment. Hu lymphoblastoid IFN, as well as hu IFN-beta and IFN-gamma can induce blast transformation and plasmacytoid differentiation in chronic B-lymphocytic leukemia cells, which is another indication that abrogation of maturation arrest contributes to the IFN-induced remissions obtained in some patients with B-cell malignancies. It is important to stress that such effects are obtained with cells from some, but not all patients.

Interesting observations come from treatment of hairy cell leukemia patients, who have a form of B cell-derived leukemia that is successfully treated with IFN-alpha. In many patients the IFN-alpha treatment results in the disappearance of hairy cells from the peripheral blood, sometimes also from the bone marrow, and restores the normal levels of platelets, granulocytes, monocytes and hemoglobin. This effect has been ascribed to the capacity of Hu-IFN-alpha or -beta but not -gamma to stimulate the lymphomyeloid stem cells of these patients toward the myelomonocytic lineage, thereby reducing the excessive formation of partially mature B cells with the phenotype of hairy cells (Quesada et al., 1984; Michalevicz and Revel 1987). Of interest is the possible involvement of co-factors, i.e. B-cell growth factor, in further enhancing IFN-induced differentiation in hairy cell leukemia (Gressler et al., 1989).

As before, the differentiation-stimulating activities of IFNs are theoretically relevant to their antitumor action, since, by inducing differentiation, they redirect cells toward normality. More recently it has been reported (Riter et al., 1992) that combination of IFN-a treatment and 2-chlorodeoxyadenosine (a purine nucleotide effective in some malignant disorders of lymphoid tissue) results in an additive antiproliferative/cytotoxic effect on hairy cell leukemia-like cell line.

1. Effects of IFN administration to Friend erythroleukemia cells induced to differentiate

Evidence of the effect of IFN on erythropoietic differentiation of FLC was a contribution of our group (Rossi et al., 1977; Dolei et al., 1980), as well as of Luftig et al., (1977), when even the mere hypothesis of IFN interaction(s) with cells was strongly questioned.

Following are the most recent data from our group on this topic.

FLC are erythroid precursors blocked in their differentiation pathway at the proerythroblastic stage; treatments of these cells with dimethylsulfoxide (Me_2SO) and other inducers causes a massive shift towards the normoblast stage of erythropoiesis (Reuben et al., 1980).

Biochemical alterations accompanying Me_2SO-induced differentiation are illustrated in Table I and comprise an arrest in the cell growth with the loss or the increase of some enzymatic activities.

Table I. Biochemical alterations accompanying dimethylsulfoxide-induced differentiation

1. Delay in onset cell replication
2. Reduced rates of incorporation of labeled precursors into DNA, RNA and protein
3. Increase in :
   - Heme
   - globin mRNA
   - globin proteins
   - hemoglobin
   - delta-aminolevulate synthetase
   - carbonic anhydrase activity
   - spectrin
   - glycophorin
4. Loss of activity of enzymes involved in de novo synthesis of purines, but retention of activity of those involved in purine metabolism via the salvage pathway.

Looking at the pattern of hemoglobin production we analyzed globin mRNA levels and chains synthesis, heme production and transferrin receptor expression during either stimulation or inhibition of FLC differentiation (Battistini et al., 1991a; Battistini et al., 1991b).

The stimulatory effect of differentiation was achieved with the addition of IFNs to Me_2SO-induced FLC and the inhibitory effect was obtained treating differentiating FLC with succinylacetone (SA), an highly specific inhibitor of ALA-dehydratase and heme synthesis.

The effect of the administration of highly purified preparations of murine IFN-$alpha_1$, -$alpha_6$ and -beta to FLC induced to differentiate are shown in Table II. Murine IFN-alpha and -beta have enhancing effects on erythroid differentiation of Me_2SO-induced FLC. The degree of differentiation is determined by evaluating the percentage of benzidine positive (B^+) cells, i.e. cells synthetizing hemoglobin. On the 3rd day of culture, Me_2SO-treated FLC exhibit increases of B^+ values from 25% to 60% upon addition of 500 U/ml IFN-beta and from 25% to 45% upon treatment with the same doses of IFN-$alpha_1$ or -$alpha_6$. Dose response studies indicate that a 10-fold higher dose of IFN-$alpha_6$ is necessary to achieve the effect elicited by IFN-beta species. Conversely IFN-beta produces a more inhibitory effect on FLC growth when compared with IFN-$alpha_6$, or -$alpha_1$ (Battistini et al., 1991b). If SA is added to Me_2SO induced FLC, a dramatic reduction of B^+ cells is observed. These effects on hemoglobin production are associated with changes in the globin chains synthesis and in the steady-state levels of the correspondent messenger RNAs (Battistini et al., 1991a).

It is known that differentiation of Me_2SO-treated FLC is associated with an increase of globin mRNAs and globin chains (Charnay et al., 1984; Profous et al., 1983). When either IFN-beta (500 U/ml) or IFN-$alpha_6$ or -$alpha_4$ are added to differentiating FLC a marked stimulation of mRNA synthesis is observed. Analysis of globin chains synthesis provides comparable results in that the level of production of all globin chains is 2-3 fold higher upon treatment with Me_2SO + IFN-beta and -alpha than with Me_2SO alone. Moreover, the ratio between the beta-major and beta-minor chain, which is 3:1 in differentiating FLC, remains essentially unmodified during IFN-enhanced differentiation (Battistini et al., 1991b).

2. Mechanisms underlying IFN-induced differentiation of FLC

Since heme plays a critical regulatory role in various metabolic processes in erythroid cells, we measured heme synthesis to assess whether the stimulation of Hb accumulation in FLC treated with Me_2SO + IFN-alpha or -beta is caused by an increased heme synthesis. The results obtained indicate that in Me_2SO-induced FLC, 500 U/ml of IFN-beta double the heme content (Table III). Thus, heme may represent a major factor mediating the correspondent marked increase of Hb synthesis induced by IFNs. In accord with this hypothesis, inhibition of heme synthesis by SA drastically decreases the levels of globin mRNAs and chain synthesis in Me_2SO-IFN treated FLC.

Heme has long been known as the prosthetic group of hemoproteins and is also involved in the regulation of the biosynthesis of many of these proteins by exerting its effects at different steps as transcription, translation, transport, assembly and protein degradation. Moreover, hemin, iron and protoporphyrin IX represent the main molecules involved in the control of the transferrin/ferritin system. Transferrin is the major and perhaps only source of iron for heme synthesis in erythroid cells (Jandl et al., 1959). The binding of diferric transferrin to its receptors on the cell surface is the first step in iron transport across the plasma membrane (Jandl et al., 1959; Morgan, 1964). Induction of FLC increases the number of transferrin receptors (TFRs) with respect to mouse reticulocytes (4.4 x 10^5/cell versus 0.86 x 10^5/cell) (Yeoh and Morgan, 1979), thereby presenting an improved model for the study of transferrin receptor synthesis regulation.

To investigate whether the biological effects elicited by IFNs on heme and hemoglobin synthesis correlate with a modulation of TfRs number, analysis of transferrin binding were performed. Previous results on the regulation of TfRs in fibroblast (Ward et al., 1982), leukemic cells (Louache et al., 1984), and mitogen activated T-lymphocytes (Pelosi et al., 1986), indicated that intracellular iron concentration modulates the number of TfRs via a negative feedback.

TABLE II

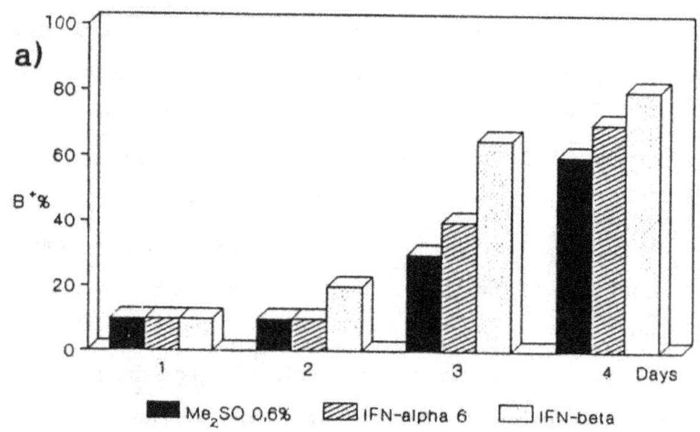

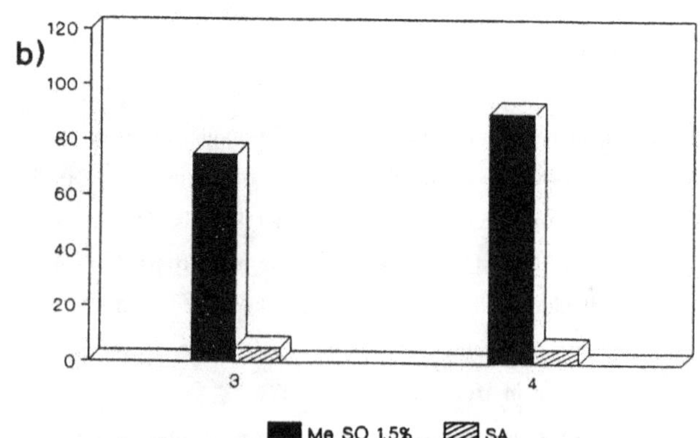

The percentage of B^+ was evaluated according to Orkin et al., 1976.

Our results indicate that in differentiating FLC, an heme-induced up-regulation of TfRs is operative instead. In fact, on day 3 and 4 of culture transferrin binding remains at high levels in Me_2SO-treated FLC undergoing erythroid differentiation whereas in control cultures it gradually declines down to the initial levels. The addition of IFN-beta (500 U/ml), IFN-$alpha_4$ or -$alpha_6$ (2000 U/ml) together with Me_2SO induces significantly higher transferrin binding with respect to control cells grown with Me_2SO alone (Table III).

The hypothesis that these effects may be mediated by the increase in heme content induced by IFNs is confirmed by experiments of inhibition of heme synthesis. The effects of the specific inhibitor of ALA-dehydratase, succinylacetone (SA), on heme synthesis and on TfRs number are shown (Table III). The addition of 1 mmol/l of SA to the cultures treated with Me_2SO induces a drastic reduction of both heme content and TfRs number particularly marked on day 4 of culture.

Table III

Cells and treatment	Transferrin receptors/cell	Heme nmole/10^6 cells
	Day 3	
745A	3.08×10^5	0.02
745A + Me_2SO (0.6%)	3.39×10^5	0.15
745A + Me_2SO (0.6%) IFN-beta	9.2×10^5	0.3
745A + Me_2SO (0.6%) + IFN-$alpha_6$	9.2×10^5	0.2
	Day 4	
745A + Me_2SO (1.5%)	4.30×10^5	0.3
745A + Me_2SO (1.5%) + SA (1mM)	2.00×10^5	0.06

The numbers of transferrin receptors was determined by Scatchard analysis (Schatchard, 1949). Heme concentration was determined fluorimetrically by a modification of the method described in Sassa et al., 1976.

Again this effect of inhibition or stimulation of TfRs number is dependent upon erythroid differentiation. In fact, the addition of hemin to control FLC not induced to differentiate elicits the down-regulation of TfRs expression (not shown) as observed in several other cell types (Ward et al., 1982; Louache et al., 1984; Pelosi et al., 1986).

In conclusion, these results indicate that in erythropoietic differentiation the increase in heme content plays a key role not only in stimulating globin chain synthesis but also in mediating optimal expression of TfRs, which are in turn necessary to sustain an optimal synthesis of heme itself; the coordinate regulation of these events by intracellular heme may thus represent a key mechanism in the control of hemoglobin synthesis.

SIGNAL TRANSDUCTION OF DIFFERENTIATION CONTROL

The initial event in IFN action appears to be the interaction with its cell surface receptor (Pestka et al., 1987). Following this interaction, the nature of the signal transduction mechanism involved in mediating the effects of IFN is not clear. Possible candidates for transducing receptor signals to the IFN-mediated differentiation control pathways are members of a family of phospholipid-dependent serine/threonine specific kinases, known as protein kinase C (PKC) (Nishizuka, 1986). There is indirect and direct evidence suggesting activation of PKC by IFNs (Akai and Larner 1990; Pfeffer et al., 1991a; Reich and Pfeffer, 1990; Tiefenbrun and Kimchi, 1991; Lanciotti et al., 1989). A rapid and transient dose-dependent rise in diacylglycerol concentration was reported after exposure of primary human fibroblasts to HuIFNs-alpha,beta and gamma (Fan et al., 1988). Protein kinase inhibitors were used in some systems to correlate PKC activation with the transcriptional induction of certain genes and the antiviral effects of IFN (Pfeffer et al., 1991a; Lin et al., 1991).

However, further work is now directed to study whether PKC is coupled to molecular events that lead to growth inhibition and differentiation control. The ability of IFN to suppress cellular growth has been shown to correlate in specific cell systems with the induction of differentiation (Rossi, 1985; Rossi et al., 1987; De Maeyer and De Maeyer-Guignard, 1988; Romeo et al., 1989). As noted previously, in human melanoma cell cultures, the combination of IFN-beta and mezerein results in an induction of terminal differentiation with a concomitant loss of proliferative capacity and an increase in the synthesis of melanin (Ahmed et al., 1989; Graham et al., 1991). Employing the PKC inhibitor [1-(5-isoquinolinylsulfonyl)-2-methylpiperazine dihydrochloride] (H-7), mezerein-induced growth suppression could be partially reversed, whereas the induction of a loss of proliferative potential by the combination of IFN-beta plus mezerein was not modified by H-7 treatment (Guarini et al., 1990). These observations suggest that the induction of terminal differentiation in human melanoma cells by the combination of IFN-beta plus mezerein may be mediated by cellular changes other than PKC activation. In line with these data, it was previously reported that phosphatidylinositol turnover is not a general regulator of neuroblastoma cell differentiation (Lanciotti et al., 1989).

In addition, the conventional approach of depleting cells from PKC activity has been undertaken by Tiefenburn and Kimchi (Tiefenbrun et al., 1991) to investigate whether PKC mediates some of the growth inhibitory responses to IFN. In two different hematopoietic cell lines, the authors reported that the cell cycle responses to IFN-alpha or -beta (i.e., the G_0/G_1 arrest induced by IFN) have been completely abrogated in PKC depleted cells.

It has been reported previously that PKC mediates opposing signals such as growth stimulation or inhibition and cellular maturation and differentiation (Vandenbark and Niedel, 1984). A possible explanation for the diverse and antagonistic roles of PKC in cellular growth control may be that, in each cell type and at each stage of maturation, the downstream targets of the kinase are completely contribute to these pleiotropic effects due to selective stimulation of different PKC isozymes by different agonists (Pfeffer et al., 1991b).

Additional early events associated with IFN action, following receptor binding, include the activation of specific transcriptional changes in target cells. The gene(s) that is stimulated by IFN possesses an ISRE that functions as an inducible enhancer for transcriptional activation by IFN (Cohen et al., 1988; Rutherford et al., 1988). The initiation of transcription by IFN appears to involve the activation of latent transcription binding factors localized in the cytoplasm of the target cell and the subsequent translocation of these factors to the nucleus (Dale et al., 1989). Recent studies of the effects of PKC inhibitors on the expression of IFN-responsive genes provide further support for a link between PKC activation and early cellular responses to IFN-alpha (Reich and Pfeffer, 1990). As indicated above, treatment of HeLa cells with IFN-alpha results in a rapid increase in PKC, as indicated by an increased binding of [^{3}H]phorbol 12,13-dibutyrate to intact cells (Reich and Pfeffer, 1990). PKC inhibitors, specifically H-7 and staurosporine, block the ability of IFN-alpha to induce an antiviral state against vesicular stomatitis virus infection in HeLa cells (Reich and Pfeffer, 1990). In addition, both H-7 and staurosporine inhibit the activation of the IFN-stimulated gene(s) (ISG), ISG-15 and ISG-54, in HeLa cells treated with IFN-alpha (Reich and Pfeffer, 1990). The inhibition of ISG-15 and ISG-54 transcription by the PKC inhibitors was found to correlate with the failure of IFN-alpha to induce the appearance of several DNA-binding factors that specifically recognize the ISRE (Reich and Pfeffer, 1990).

In the human melanoma cell line H0-1, mezerein, a PKC activator, also failed to induce the expression of ISG-15 and ISG-54, whereas IFN-beta induced expression of both these ISGs (Ahmed et al., 1989). On the other hand, it has been previously reported (Levy et al., 1989) that the conventional receptor-mediating signaling mechanisms that involve PKC, cyclic adenosine-monophosphate (cAMP)-dependent protein kinase A or fluxes in intracellular pH or calcium, do not function in IFN-induced activation of ISRE-directed gene expression (Levy et al., 1989).

Thus a role for kinases in IFN-alpha and IFN-gamma signal trasduction pathways has been suggested previously on the basis of differential effects of kinase inhibitors on the trascriptional activation of IFN-inducible genes or on the activation of specific DNA-binding factors. In particular, for IFN-alpha,beta the involvement of specific isoforms of PKC has been proposed although agonists of PKC activity do not mimic IFN action and PKC down regulation does not alter the transcriptional response to IFN. More recently, on the other hand, staurosporine and K-252a have been shown to prevent activation of the signaling pathway, even in the absence of PKC (Kessler and Levy, 1991). The involvement of a novel protein tyrosine kinase in the current model of the biochemical mechanisms of IFN-alpha,beta action has been proposed (Velazquez et al., 1992). This tyrosine kinase appears to link

the IFN-alpha,beta receptor to the cytoplasmic transcription factors that mediate activation of ISGs.

Stimulation of phospholipase A_2 (that is a source of second messenger for receptor-mediating signaling) and the consequent release of arachidonic acid from membrane phospholipid have also been studied as possible molecular mechanisms in signaling IFN-alpha specific growth inhibitory response through activation of cytoplasmic ISRE-trans acting factors (Hannigan and Williams, 1991).

As is the case with phospholipids, sphingolipids may act as a precursor for active metabolites which are generated in response to cell agonists. In fact, the effects of IFN-gamma and tumor necrosis factor (TNF-alpha) on sphingomyelin turnover have been proven in HL-60 cells which have been induced to differentiate into monocyte-like cells by treatment with IFN-gamma or TNF-alpha (Kim et al., 1991). These data suggest that sphingomyelin turnover may be an important signaling mechanism for the action of IFN-gamma and tumor necrosis factor-alpha on cell differentiation. In addition, because of the prolonged duration of occurrence of the sphingomyelin cycle with respect to the phosphatidylinositol cycle, the authors hypothesize that sphingomyelin turnover may be better suited for mediating long-term cellular responses, such as differentiation, as compared with the more prompt responses that have been so far associated with phosphatidylinositol turnover.

<u>Protein binding to IFN response enhancer of 2-5A synthetase in FLC</u>.

2-5A synthetase, like all the IFN induced genes, contains a transcriptional enhancer element designated the IFN stimulated response element (ISRE) (Benech et al., 1987; Hug et al., 1988). This element is able to confer IFN-alpha,beta and IFN-gamma inducibility to a reporter gene (Cohen et al., 1988). The <u>in vitro</u> formation of protein-DNA complexes with the ISRE of the 2-5A synthetase or related sequences, seems to coincide with <u>in vivo</u> induction of the corresponding genes in response to IFNs (Dale et al., 1989; Levy et al., 1989). We studied the mechanism involved in the induction of transcription of the 2-5A synthetase gene by type I (alpha,beta) and type II (gamma) interferons. In order to dissect the pathways of induction of the two types of IFNs, we studied the formation of protein-ISRE complexes both in wild type and in interferon-resistant FLCs (Coccia et al., 1991). These FLC variants are totally resistant to type I IFN and sensitive to type II (3CL8 cells) or resistant to both types of IFNs (3gammaR8). This resistance is not mediated by changes in the IFN receptor number or affinity (Coccia et al., 1988); therefore the resistant cell clones appear to be deficient in intracellular factors which transduce the IFN signal.

We showed that in w.t. FLC, one ISRE-protein complex, designated F_1, is induced following IFN-alpha,beta treatment whereas two ISRE-protein complexes Fg and F_1 are formed following IFN-gamma treatment. Both type I and II IFNs fail to induce 2-5A synthetase mRNA transcription in FLC resistant to both IFN types, accordingly no IFN-specific ISRE-protein complexes are detected. Conversely, the presence of only one specific ISRE complex, Fg, correlates with the transcription of the mRNA and the appearance in the cytoplasm of the functional protein after IFN-gamma treatment in type I IFN-resistant cells. These findings suggest that F_1 and Fg represent two distinct effector complexes by which type I and type II IFNs, respectively, induce 2-5A synthetase (Coccia et al., 1991).

IFN-INDUCIBLE ENZYMES IN THE REGULATION OF CELL DIFFERENTIATION.

At least some of the regulatory functions attributed to IFN are mediated through IFN-induced enzymes, generally referred to as the IFN system. This IFN system includes 2-5A synthetase, 2-5A-activated RNase and the double stranded RNA (dsRNA) activated protein kinase. The first enzyme catalyzes the synthesis of oligomers of adenylic acid with 2'-5' phosphodiester bonds (Kerr and Brown, 1978), which, in turn, activate a latent endonuclease resulting in the degradation of messenger and ribosomal RNA molecules (Williams et al., 1978). The dsRNA-activated protein kinase phosphorylates the alpha subunit of initiation factor eIF-2. Activation of 2-5A system and protein kinase may, therefore, directly cause inhibition of protein synthesis (Kerr and Brown, 1978; Williams et al., 1978; Baglioni, 1979), as manifested by antiproliferative effects. The appearance of 2-5A is considered a valuable marker for the cellular response to IFN action. Measurements of 2-5A synthetase activity indicate that the concentration of this enzyme rises during maturation of T lymphocytes, erythroid cells and monocytes.

Thus, 2-5A synthetase per se may be involved in growth control. Furthermore, G_0-arrested cells have higher concentrations of 2-5A synthetase in the nucleus than are found in growing cells. Exogenous nonphosphorylated 2-5A has been reported to inhibit the mitogenic response of mouse lymphocytes and BALB/c 3T3 cells (Revel et al., 1984). Other connections between 2-5A synthetase and growth have been discovered. For example, stimulation of growth of human diploid fibroblasts by EGF increases synthetase levels. PDGF, besides inducing c-myc and c-fos (and IFN-beta) and enabling G0 cells to enter G1, has been shown to induce 2-5A synthetase RNA in 3T3 cells (Zullo et al., 1985). Thus, studies of cells under different growth conditions indicate that 2-5A synthetase activity is greater in confluent and serum-starved cells than in proliferating cells, even in the absence of IFN. Conversely, there are indications that IFN can inhibit proliferation in the absence of any 2-5A synthetase activity. Thus, the involvement of the enzyme on growth control remains controversial. The levels of 2-5A-dependent ribonuclease are higher when cells have reached confluence. In this case, however, the changes are quantitatively smaller (see Romeo et al., 1989).

2-5A synthetase appearance and induction in chicken embryo erythrocytes during development have been observed by Sokawa and Sokawa (Sokawa and Sokawa, 1986). Enzyme activity first appears in the embryos on the 15th day of incubation; a marked increase is seen 1 or 2 days after hatching. The involvement of 2-5A synthetase in the differentiation of neocartilage in perinatal mice has also been reported (Maor et al., 1990). It has become evident that the activity of 2-5A synthetase is indeed different in cellular compartments that are at various stages of differentiation. In the neonatal condyle, the highest level of activity is encountered in proliferating and as yet undifferentiated prechondrocytes, whereas fully differentiated chondrocytes show a marked decrease in the activity of this enzyme.

The activation of the IFN system during myogenesis in vitro has also been demonstrated (Birnbaum et al., 1990). The activity of 2-5A synthetase and dsRNA-activated protein kinase and the expression of 2-5A synthetase coding genes have been examined during myogenesis. It has been demonstrated that the activity of the enzymes is transiently increased in cultured myoblasts, reaching a peak activity on the 3rd day in culture and then declining to a basal level. The same kinetics of 2-5A synthetase activity is evident in myoblasts from chick, rat or mouse origin.

With respect to the other well-known system induced by IFNs, the dsRNA-dependent protein kinase pathway, a physiological role has been demonstrated for this enzyme, which appears to regulate growth and adipose conversion of 3T3-F442A cells (Petryshyn et al., 1988). In particular, IFN has an autocrine effect on these cells that leads to transient expression of the kinase activity at specific stages of growth and differentiation. More recently, the same authors have reported that the same cells, when cultured under conditions which are nonpermissive for differentiation, exhibit significantly reduced dsRNA-dependent protein kinase activity. Further, this reduction is due to the presence of elevated levels of a novel inhibitor of dsRNA-dependent protein kinase activation. This inhibitor is a physiologic regulator of dsRNA-dependent protein kinase, since its activity correlates with the ability of 3T3-F442A cells to undergo adipose conversion (Judware and Petryshyn, 1991). Total cytoplasmic RNA from 3T3-F442A cells contains a regulatory RNA(s) capable of activating dsRNA-dependent eIF-2alpha kinase. The activation of the dsRNA-dependent eIF-2alpha kinase by regulatory RNA is prevented by addition of a high concentration of poly(I)-poly(C) (Li and Petryshyn, 1991).

It appears now that the involvement of the IFN system in the differentiation process is not limited to certain hematopoietic cells, but has a much wider significance. The exact role that the IFN system may fulfill during differentiation remains unclear.

However, another consideration may be that nuclear protooncogenes, particularly c-myc, which are essential for cell-growth, are switched off during the differentiation of several cell systems, including hematopoietic cells (Gonda and Metcalf, 1984). In some cases, the expression of c-myc is regulated post-transcriptionally such as in the murine embryonal carcinoma cell line F9 which is induced to differentiate with retinoic acid and cyclic AMP (Dean et al., 1986). One interpretation is that the stability of c-myc mRNA transcripts is reduced in this system, which could certainly be the result of nuclease activity induced by 2-5A oligomers, the enzymatic products of 2-5A synthetase (Baglioni, 1979). The reduction in c-myc expression has been demonstrated also during differentiation of myoblasts (Olson et al., 1987), and therefore a similar mechanism in these cells is possible. Although the discussion above has focused briefly on IFN effects on the representative growth-related cellular gene c-myc, parallel modulatory effects will undoubtedly be evident when other gene(s) involved in the regulation of cellular proliferation and differentiation are examined.

These reports suggest that the IFN system may be involved in the differentiation process of cells of various lineages and may indeed be crucial for the normal progression of differentiation. An altered expression of the IFN system or a lack of response to its signals may therefore lead to the development of malignancy.

ACKNOWLEDGEMENTS

This work was supported by grants from Associazione Italiana Ricerca sul Cancro, the Italy-USA Program on Therapy of Tumors, the AIDS project of the Ministry of Public Health, Istituto Superiore di Sanità, Rome and Consiglio Nazionale delle Ricerche, Progetto Finalizzato ACRO, codice pubblicazione n° 132. The skillful secretarial assistance of Mmes. S. Tocchio and O. Fantauzzo is gratefully acknowledged.

REFERENCES

Ahmed MA, Vita JR, Ferrone S, Fisher PB (1989) Induction of Differentiation in human melanoma cells by the combination of different classes of interferons or interferon plus merezein. Diamond L, Wolman SR, (eds). Viral Oncogenesis and Cell Differentiation. N.Y. Acad. Sci., New York, 567: 328-333.

Akai H, Larner A. (1990) Characterization of two independent mechanisms by which interferon-induced gene expression is down regulated. Cell Regul 1: 707-713.

Andre P, Braun S, Passaquin AC, Coupin G, Bartholeyns J, Warter JW, Poindron P (1988). Rat interferon enhances the expression of acetylcholine receptors in rat myotubes in culture. J Neurosci Res 19: 297-302.

Baglioni C (1979) IF-induced enzymatic activities and their role in the antiviral state. Cell 17: 255-264.

Battistini A, Coccia EM, Bulgarini D, Scalzo S, Fiorucci G, Romeo G, Affabris E, Testa U, Rossi GB, Peschle C (1991a) Intracellular heme coordinately modulates globin chains synthesis, transferrin receptor number and ferritin content in differentiating Friend erythroleukemia cells. Blood 78: 2098-2103.

Battistini A, Marziali G, Albertini R, Habetswallner D, Bulgarini D, Coccia EM, Fiorucci G, Romeo G, Orsatti R, Testa U, Affabris E, Peschle C, Rossi GB (1991b) Positive modulation of hemoglobin, heme and transferrin receptor synthesis by murine interferon-alpha and -beta in differentiating Friend cells: pivotal role of heme synthesis. J Biol Chem 266: 528-535.

Benech P, Vigneron M, Peretz D, Revel M, Chebath J (1987) Interferon-responsive regulatory elements in the promoter of the human 2'-5'-oligo (A) synthetase gene. Mol Cell Biol 7: 4498-4504.

Birnbaum M, Trink B, Shainberg A, Salzberg S (1990) Activation of the interferon system during myogenesis <u>in vitro</u>. Differentiation 45: 138-145.

Charnay P, Treisman R, Mellon O, Chao M, Axel R, Maniatis T (1984) Differences in human alpha- and -beta globin gene expression in mouse erythroleukemia cells: the role of intragenic sequences. Cell 38: 251-263.

Cioè L, O'Brien TG, Diamond L (1980) Inhibition of adipose conversion of Balb/c 3T3 cells by interferon and 12-0-tetradecanoylphorbol-13-acetate. Cell Biol Int Reports 4: 255.

Coccia EM, Federico M, Romeo G, Affabris E, Cofano F, Rossi GB (1988) Interferon-alpha/beta and -gamma resistant Friend cell variants exhibiting receptor sites for interferons but no induction of 2-5A synthetase and 67K protein kinase. J. Interferon Res. 7: 113-127.

Coccia EM, Vaiman D, Raber J, Marziali G, Fiorucci G, Orsatti R, Cohen B, Nissim N, Romeo G, Affabris E, Chebath J, Battistini A (1991) Protein binding to the interferon response enhancer correlates with interferon-induction of 2-5A synthetase in normal and interferon-resistant Friend cells. J Virol 65: 2081-2087, 1991.

Cohen B, Peretz D, Vaiman D, Benech P, Chebath P (1988) Enhancer-like interferon responsive sequences of the human and murine (2'-5') oligo (A) synthetase gene promoters. EMBO J. 7: 1411-1419.

Cohen B, Gothelf Y, Vaiman D, Chen L, Revel M, Chebath J (1991) Interleukin-6 induces the (2'-5') oligoadenylate synthetase gene in M1 cells through an effect on the interferon-responsive enhancer. Cytokine 3: 83-91.

Dale TC, Imam AMA, Kerr IM, Stark GR (1989) Rapid activation by interferon-alpha of a latent DNA-binding protein present in the cytoplasm of untreated cells. Proc Natl Acad Sci USA 86: 1203-1207.

Dale TC, Rosen JM, Guille MJ, Lewin AR, Porter AG, Kerr IM, Stark GR (1989) Overlapping sites for constitutive and induced DNA binding factors involved in interferon-stimulated transcription. EMBO J. 8: 831-839.

Dean M, Levine RA, Campisi J (1986) C-myc regulation during retinoic acid-induced differentiation of F-9 cells is posttranscriptional and associated with growth arrest. Mol Cell Biol 6: 518-524.

De Maeyer E, De Maeyer-Guignard (eds) (1988) Interferon and other regulatory cytokines. Wiley Interscience Publication New York.

Dolei A, Colletta G, Capobianchi MR, Rossi GB, Vecchio G (1980) Interferon effects on Friend leukemia-cells. I. Expression of virus and erythroid markers in untreated and dimethyl sulphoxide-treated cells. J Gen Virol 46: 227-236.

Exley R, Nathan P, Walker L, Gordon J, Clemens MJ (1987) Anti-proliferative effects of interferons on Daudi Burkitt-lymphoma cells: Induction of cell differentiation and loss of response to autocrine growth factors. Int J Cancer 40: 53-57.

Fan X-D, Goldberg M, Bloom BR (1988) Interferon-gamma-induced transcriptional activation is mediated by protein kinase C. Proc Natl Acad Sci USA 85: 5122-5125.

Fisher PB, Miranda AF, Babiss LE, Pestka S, Weinstein B (1983) Opposing effects of interferon produced in bacteria and of tumor promoters on myogenesis in human myoblast cultures. Proc Natl Acad Sci USA 80: 2961-2965.

Gonda TJ, Metcalf D (1984) Expression of myb, myc and fos proto-oncogenes during the differentiation of a murine myeloid leukemia. Nature 310: 249-251.

Graham GM, Guarini L, Moulton TM, Datta S, Ferrone S, Giacomini P, Kerbel RS, Fisher PB (1991) Potentiation of growth suppression and modulation of the antigenic phenotype in human melanoma cells by the combination of recombinant human fibroblast and immune interferons. Cancer Immunol Immunother, 32: 382-390.

Gressler VH, Weinkauff RE, Franklin WA, Golomb HM (1989) Is there a direct differentiation-inducing effect of human recombinant interferon on hairy cell leukemia in vitro? Cancer 64: 374-378.

Guarini L, Graham GM, Datta S, Kelly C, Ferrone S, Fisher PB (1990) IV Confer. Different. Ther., Chiba, Japan, Abst. 96.

Hannigan GE, Williams RG (1991) Signal transduction by interferon-alpha through arachidonic acid metabolism. Science 251: 204-207.

Harada H, Willison K, Sakakibara J, Miyamoto M, Fujita T, Taniguchi T (1990) Absence of the type I interferon system in EC cells: transcriptional activator (IRF-1) and repressor (IRF-2) genes are developmentally regulated. Cell 63: 303-312.

Hattori T, Pack M, Bougnoux PH, Chang ZL, Hoffman T (1983) Interferon-induced differentiation of U937 cells. Comparison with other agents that promote differentiation of human myeloid or monocyte-like cell lines. J Clin Invest 72: 237-244.

Hug H, Costas M, Staeheli P, Aebi M, Weissman C (1988) Organization of the murine Mx gene and characterization of its interferon-and virus-inducible promoter. Mol Cell Biol 8: 3065-3079.

Improta T, Salvatore AM, Di Luzio A, Romeo G, Coccia EM, Calissano P (1988) IFN-gamma facilitates NGF-induced neuronal differentiation in PC12 cells. Exp Cell Res 179: 1-8.

Jandl JH, Inman JK, Simmons RL, Allen DW (1959) Transfer of iron from serum iron-binding protein to human reticulocytes. J Clin Invest 38: 161-185.

Jiang H, Su Zz, Boyd J, Fisher PB (1992) Multistep model of terminal differentiation in human melanoma cells treated with the combination of recombinant human fibroblast interferon (IFN-beta) and (MEZ). Biomed. & Pharmacother. 46: 315-319.

Judware R, Petryshyn R (1991) Partial characterization of a cellular factor that regulates the double-stranded RNA-dependent eIF-2alpha kinase in 3T3-F442A fibroblasts. Mol Cell Biol 11: 3259-3267.

Keay S and Grossberg SE (1980) Interferon inhibits the conversion of 3T3-L1 mouse fibroblasts into adipocytes. Proc Natl Acad Sci USA 77: 4099.

Kerr IM, Brown RE (1978) pppA2-5A2'p5'A: Inhibitor of protein synthesis synthesized with an enzyme fraction from interferon-treated cells. Proc Natl Acad Sci USA 75: 256-260.

Kessler DS, Levy DE (1991) Protein kinase activity required for an early step in interferon-alpha signaling. J. Biol. Chem. 266: 23471-23476.

Kim M-Y, Linardic C, Obeid L, Hannun Y (1991) Identification of sphingomyelin turnover as an effector mechanism for the action of tumor necrosis factor alpha and gamma-interferon. J Biol Chem 266: 484-489.

Lanciotti M, Cornaglia-Ferraris P, Ponzoni M (1989) Phosphatidylinositol turnover is not a general regulator of neuroblastoma cell differentiation: comparison between two differentiating agents, retinoic acid and gamma-interferon. Febs Letters 243: 285-288.

Levitt M, Nair S, Mayotte J (1990) Relation of beta interferon to growth and squamous differentiation in human lung cancer cell lines. IV Confer Different Ther, Chiba, Japan, Abst. 98.

Levy DE, Kessler DS, Pine R, Darnell JE Jr (1989) Cytoplasmic activation of ISGF3, the positive regulator of interferon-a-stimulated transcription, reconstituted in vitro. Genes Dev. 3: 1362-1371.

Li J, Petryshyn A (1991) Activation of the double-stranded RNA-dependent eIF-2alpha kinase by cellular RNA from 3T3-F442A cells. Eur J Biochem 195: 41-48.

Lin H-Y, Davis PJ, Chadha KC, Thacore HR (1991) Interferon mediated cellular antiviral state: different signal transduction pathways for interferon-alpha/beta and interferon-gamma in human and murine cells. Annual Meeting of the ISIR, Nice, France, p. S212.

Lippman SM, Parkinson DR, Itri LM, Weber RS, Schantz SP, Ota DM, Schusterman MA, Krakoff IH, Gutterman JU, Hong WK (1992) 13-cis-retinoic acid and interferon-alpha-2a: effective combination therapy for advanced squamous cell carcinoma of the skin. J Natl Cancer Inst 84: 235-245, 1992.

Louache F, Testa U, Pelicci P, Thomopoulos P, Titeux M, Rochant H (1984) Regulation of transferrin receptors in human hematopoietic cell lines. J Biol Chem 259: 11576-11582.

Luftig RB, Coscience JF, Skoultchi A, McMillan P, Revel M, Ruddle FH (1977) Effect of interferon on dimethyl sulfoxide-stimulated Friend erythroleukemic cells: ultrastructural and biochemical study. J Virol 23: 799-810.

Maciejewski J, Wachter M, Elstner E, Weber H, Diezel W, Seiler FR, Volk HD (1990) Cytofluorometric and cytomorphologic analysis of human bone marrow cells derived from stromal cultures stimulated by granulocyte-macrophage colony-stimulating factor, interferon-gamma and splenopentin. Eur J Immunol 20: 1209-1213.

Maor G, Salzberg S, Silbermann M (1990) The activity of 2,5-oligoadenylate synthetase, an interferon-induced enzyme, is coupled to the differentiation state of mouse condylar cartilage. Differentiation 44: 18-24.

Michalevicz R, Revel M (1987) Interferons regulate the in vitro differentiation of multilineage lympho-myeloid stem cells in hairy cell leukemia. Proc Natl Acad Sci USA 84: 2307-2311.

Morgan EH (1964) The interaction between rabbit, human and rat transferrin and reticulocytes. Br J Hematol 10: 442-447.

Nakamaki T, Sakashita A, Hino K-I, Suzuki K, Tomoyasu S, Tsuruoka N, Honma Y, Hozumi M (1990) Difference in effects of interferon-alpha and interferon-gamma on the induction of differentiation of retinoic acid-treated acute myeloid leukemia cells in primary culture. Leukemia Res 14: 785-794.

Nishizuka Y (1986) Studies and perspectives of protein kinase C. Science 233: 305-312.

Oberg F, Larsson L-G, Anton R, Nilsson K (1991) Interferon gamma abrogates the differentiation block in v-myc-expressing U-937 monoblasts. Proc Natl Acad Sci USA 88: 5567-5571.

Olson EN, Spizz G, Tainsky MA (1987) The oncogenic forms of N-ras and H-ras prevent skeletal myoblast differentiation. Mol Cell Biol 7: 2104-2111.

Orkin S, Swan D, Leder P (1975) Differential expression of alpha- and -beta globin genes during differentiation of cultured erythroleukemic cells. J Biol Chem 250: 8753-8760.

Pelosi E, Testa U, Louache F, Thomopoulos P, Salvo G, Samoggia P, Peschle C (1986) Expression of transferrin receptors in phytohemagglutinin-stimulated human T-lymphocytes. Evidence for a three-step model. J Biol Chem 261: 3036-3042.

Perussia B, Dayton ET, Fanning V, Thiagarajan P, Hoxie J, Trinchieri G (1983) Immune interferon and leukocyte-conditioned medium induce normal and leukemic myeloid cells to differetiate along the monocytic pathway. J Exp Med 158: 2058-2080.

Pestka S, Langer JA, Zoon KC, Samuel CE (1987) Interferons and their actions. Ann Rev Biochem 56: 727-777.

Petryshyn R, Chen JJ, London IM (1988) Detection of activated double-stranded RNA-dependent protein kinase in 3T3-F442A cells. Proc Natl Acad Sci USA 85: 1427-1431.

Pfeffer LM, Eisenkraft BL, Reich NC, Improta T, Baxter G, Daniel-Issakani S, Strulovivi B (1991a) Transmembrane signaling by interferon-alpha involves diacylglycerol production and activation of the epsilon isoform of protein kinase C in Daudi cells. Proc Natl Acad Sci USA 88: 7988-7992.

Pfeffer LM, Reich NC, Strulovici B (1991b) Transmembrane signaling by interferon-alpha involves diacylglycerol production and activation of protein kinase C-epsilon in the Daudi line of human lymphoblastoid cells. Annual Meeting of the ISIR, Nice, France, p. S98.

Profous-Juchelka HR, Reuben RC, Marks PA, Rifkind RA (1983) Transcriptional and post-transcriptional regulation of globin gene accumulation in murine erythroleukemia cells. Mol Cell Biol 3: 229-232.

Quesada JR, Reuben J, Manning JT, Hersh EM, Gutterman JU (1984) Alpha-interferon for induction of remission in hairy cell leukemia. N Engl J Med 310: 15-18.

Reich NC, Pfeffer LM (1990) Evidence for involvement of protein kinase C in the cellular response to interferon-alpha. Proc Natl Acad Sci USA 87: 8761-8765.

Reuben RC, Rifkind RA, Marks PA (1980) Chemically induced murine erythroleukemic differentiation. Biochim Biophys Acta 605: 325-346.

Revel M (1984) In: Antiviral drugs and interferon: the Molecular basis of their activity (Becker, Y, ed.) pp. 375-433, Martinus Nijhoff.

Riter Z, Tomson S, Taylor MW (1992) The effects of combination treatment of interferon-a and 2-chlorodeoxyadenosine on hairy-cell leukemia-like cell line. Biom & Pharmacother 46: 317-321.

Romeo G, Fiorucci G, Rossi GB (1989) Interferons in cell growth and development. Trends in Genetics 5: 19-24.

Rossi GB, Dolei A, Cioè L, Benedetto A, Matarese GP, Belardelli F, Rita G (1977) Interferon and Friend leukemia cells in culture. "Texas Rep. on Biol. and Med." 35: 420-428.

Rossi GB (1985) Interferon and cell differentiation. Interferon, Academic Press New York 6: 31-68.

Rossi GB, Affabris E, Romeo G, Federico M, Coccia EM (1987) Interferon effects on cell differentiation. In: The interferon System: A Current Review, Dianzani F, Baron J, Stanton GJ, Fleischmann WR Jr (eds) The University of Texas Press, Austin, Texas, pp. 285-297.

Rutherford MN, Hannigan GE, Williams BRG (1988) interferon-induced binding of nuclear factors to promoter elements of the 2-5A synthetase gene. EMBO J 7: 751-759.

Saneto R and Johnson HM (1982) Interferon inhibition of hexose monophosphate Shunt activity as the mechanism of blocking differentiation of preadipose cell line. Bioch Biophys Res Comm 106: 373.

Sassa S (1976) Sequential induction of heme pathway enzymes during erythroid differentiation of mouse Friend leukemia virus-infected cells. J Exp Med 143: 305-315.

Scatchard G (1949) The attraction of proteins for small molecules and ions. Ann N Y Acad Sci 51: 660-672.

Sokawa J, Sokawa Y (1986) (2'-5')oligoadenylate synthetase in chicken embryo erythrocytes and immature red blood cells. J Biochem 99: 119-124.

Testa U, Ferbus D, Gabbianelli M, Pascucci B, Boccoli G, Louache F, Thang MN (1988) Effect of endogenous and exogenous interferons on the differentiation of human monocyte cell line U937. Cancer Res 48: 82-88.

Tiefenbrun N, Kimchi A (1991) The involvement of protein kinase C in mediating growth suppressive signals of interferons in hematopoietic cells. Oncogene 6: 1001-1007.

Tomita Y, Hasegawa S (1984) Multiple effects of interferon on myogenesis in chicken myoblast cultures. Biochem Biophys Acta 804: 370-376.

Tomida M, Yamamoto Y, Hozumi M (1980) Stimulation by interferon of induction of differentiation of mouse myeloid leukemic cells. Cancer Res 40: 2919-2924.

Tomida M, Yamamoto Y, Hozumi M (1982) Stimulation by interferon of induction of differentiation of human promyelocytic leukemia cells. Biochem Biophys Res Comm 104: 30-37.

Vandenbark GR, Niedel JF. (1984) Phorbol diesters and cellular differentiation. J Natl Cancer Inst 73: 1013-1019.

Velazquez L, Fellous M, Stark GR, Pellegrini S (1992) A protein tyrosine kinase in the interferon-alpha,beta signaling pathway. Cell 70: 313-322.

Ward JH, Kushner JP, Kaplan J (1982) Transferrin receptors of human fibroblasts. Analysis of receptor properties and regulation. Biochem J 208: 19-26.

Weinberg JB, Misukonis MA, Hobbs MM, Borowitz MJ. (1986) Cooperative effects of gamma interferon and 1-alpha, 25-dihydroxyvitamin D3 in inducing differentiation of human promyelocytic leukemia (HL-60) cells. Exp Hematol 14: 138-142.

Williams BRG, Kerr IM, Gilbert CS, White CN, Ball LA (1978) Synthesis and breakdown of pppA2'5'A2'5'A and transient inhibition of protein synthesis in extracts from interferon-treated and control cells. Eur J Biochem 92: 455-462.

Yamamoto-Yamaguchi Y, Tomida M, Hozumi M (1983) Effect of mouse interferon on growth and differentiation of mouse bone marrow cells stimulated by two different types of colony-stimulating factor. Blood 62: 597-601.

Yeoh GC, Morgan EH (1979) Dimethyl sulphoxide induction of transferrin receptors on Friend erythroleukemia cells. Cell Differentiation 8: 331-343.

Zullo JN, Cochran BH, Huang AS, Stiles CD (1985) Platelet-derived growth factor and double-stranded ribonucleic acids stimulate expression of the same genes in 3T3 cells. Cell 43: 793.

INDUCTION OF TUMOR CELL DIFFERENTIATION AS A MECHANISM OF ACTION OF DNA-SPECIFIC ANTITUMOR AGENTS

Alexander Bloch
Roswell Park Cancer Institute
Elm and Carlton Streets
Buffalo, NY 14263

Since tumor cells are characterized by their capacity for sustained proliferation at a stage of incomplete maturation, induction of their differentiation accompanied by cessation of growth constitutes a potentially promising approach for therapy. Numerous DNA-specific antitumor agents such as arabinosyl cytosine, daunomycin and actinomycin D have been shown capable of inducing tumor cell maturation [1,2,4,6,8,11,13,15,16,18-30], and we have examined the mechanism by which they initiate the maturation process.

To perform the studies, we selected ML-1, a human myeloblastic leukemia cell line, derived from a patient with AML [17]. These cells proliferate in RPMI 1640 medium supplemented with 10% fetal bovine serum (FBS) with a doubling time of approximately 30 hours [25]. Under these conditions, only a small fraction (<5%) of the cell population exhibits some characteristics of more mature cells. The majority of the cells proliferate at the myeloblast stage. Since both cell growth and cell differentiation are mediated by specific growth and differentiation factors, retention of the leukemic cells in their proliferation mode implies that they respond to growth but not to differentiation signals. Thus, the initial question concerned the identity of these factors and the manner in which the cells respond to their signal.

Table 1 shows the cytokines that are capable of initiating ML-1 cell growth or differentiation. Column 2 shows that, in the absence of any cytokines, 10% FBS can support full growth of the culture. Growth also takes place when IGF_1 is added to the medium in addition to FBS. But when TNF-α or TGF-β are supplied, differentiation occurs [10,27]. In the absence of FBS (Column 3), the cytokines are unable to induce either growth or differentiation unless transferrin (Tf) is also added (Column 4) [5].

NATO ASI Series, Vol. H 75
Cancer Therapy
Edited by N. D'Alessandro, E. Mihich, L. Rausa, H. Tapiero, and T. R. Tritton

Table 1 Ability of diverse cytokines to initiate the growth or differentiation of ML-1 human myeloblastic leukemia cells in the presence of RPMI 1640 medium ± fetal bovine serum (FBS).

Cytokine	RPMI 1640 Medium +FBS +Cytokine	RPMI 1640 Medium −FBS +Cytokine	RPMI 1640 Medium −FBS +Cytokine + Transferrin
None	growth	0	0
IGF_1	growth	0	growth
TNF-α	differentiation	0	differentiation
TGF-β	differentiation	0	differentiation

0 = No growth or differentiation
IGF_1 was added at 10ng/ml, TNF-α at 10U/ml, TGF-β at 60ng/ml and Tf at 0.5μg/ml.

Thus, Tf is required for induction of growth in the presence of IGF_1 and for induction of differentiation in the presence of TNF-α or TGF-β. This finding demonstrates that the cytokines determine the type of response that ensues - either growth or differentiation - whereas Tf is required for that response to be implemented. For a response to occur, the cytokines need to be administered to the cells prior to or simultaneously with Tf. If Tf is administered first, followed - after washing - by the cytokines, growth or differentiation does not occur. This sequence-dependence characterizes the cytokines as competence factors and Tf as the progression signal.

Growth of ML-1 cells is monitored by cell number, differentiation by the increase in the number of cells that express F_c receptors, reduce nitroblue tetrazolium dye to insoluble formazan, and display elevated acid phosphatase and non-specific esterase activity [25]. These functional changes are accompanied by distinct morphological alterations. In the presence of FBS, the DNA-specific agents induce the appearance of monocytes, which are also formed when FBS is replaced by TNF-α or TGF-β together with transferrin.

The results provided in Table 1 imply that FBS contains a level of cytokines that suffices to induce and maintain the growth of ML-1 cells, but is inadequate for initiating their differentiation. Both TNF-α and TGF-β are present in FBS, but at levels too low for induction of cell differentiation. If the level of differentiation-inducing cytokines is increased, the cells mature. This finding demonstrates that the arrest at the myelobast stage, which characterizes this neoplastic cell, is not due to a loss of its differentiation potential, but results from its inability to respond to the limited levels of differentiation-inducing cytokines that are present in the serum. This lack of response can result from a decrease in the sensitivity of the tumor cells to the prevailing level of differentiation factors, or from an increase in the cells responsiveness to the growth signal. In the human host, a deficiency in the level of cytokine produced would also cause the maturation arrest of these cells. Whichever the case, Table 1 shows that by increasing the concentration of TNF-α or TGF-β in the presence of FBS or transferrin, ML-1 cell differentiation can be induced.

The recognition that Tf is involved in establishing both growth and differentiation has come about only recently [5]. Its ability to carry iron required for growth had, in the past, been considered to be its major function. Therefore, its participation in initiating differentiation raised the question whether differentiation is also dependent upon iron transport. Our finding that organic Fe-complexes can replace Tf for growth but not for differentiation of ML-1 cells, paired with the observation that antibodies to the transferrin receptor can replace Tf as the progression signal for differentiation but not for growth, support the conclusion that Tf serves as an Fe-transporter for growth, but as a participant in signal transduction for differentiation.

The ability of multiple cytokines, e.g. TNF-α and TGF-β, to serve as inducers of differentiation competence, raises the question concerning the advantage such redundancy provides to the target cells. A likely answer is provided by our observation [9] that such factors can act synergistically, allowing for the sensitive amplification of the growth or differentiation signals they carry. Since different cytokines are, generally, produced by different cell types or tissues, the input derived from such multiple sources can promote the balanced function of integrated systems.

Such balance is achieved not only by the interaction of diverse cytokines, but is also brought about by the concentrations at which the cytokines are presented to the cell. Thus, at relatively low concentrations, TNF signals the differentiation of ML-1 cells to be initiated, whereas at higher levels it inhibits that process and allows growth to proceed [14]. This regulatory effect is mediated via high and low affinity TNF binding sites which, in effect, serve as feedback regulators of growth and differentiation. For example, treatment of ML-1 cells with the phorbol ester TPA converts them to macrophages, which produce TNF [7]. Low levels of TNF, produced by low numbers of mature cells, stimulate ML-1 cell maturation, whereas higher concentrations of TNF, produced by greater numbers of cells, inhibit maturation and promote growth. Regulation of the extent to which proliferation and differentiation occur within a developmental path appears, thus, to constitute a primary function of the cytokines.

Growth as well as differentation signals exert their ultimate effect at the level of gene expression. In ML-1 cells, the oncogene myb is a key determinant of growth and differentiation. When myb is expressed, growth can occur. When the differentiation inducers TNF-α, TGF-β or TPA are supplied to the cells in the presence of FBS or Tf, myb expression is rapidly suppressed, DNA synthesis is inhibited and growth ceases, followed by the initiation of differentiation [7]. Apparently, the induction of differentiation is tied to the suppression of growth-regulating genes. If this is the case, selective interference with the expression of such genes should facilitate the cells' ability to initiate differentiation. This facilitation is, in fact, achievable by the use of DNA-specific agents.

When applied in the presence of FBS, DNA-specific antitumor agents - whether inhibitors of synthesis (ara C), replication (daunomycin) or transcription (actinomycin D) - induce the differentiation of ML-1 as well as of other tumor cell types at concentrations that are minimally cytotoxic. In ML-1 cells, that induction was found to be dependent upon TNF-α and/or TGF-β together with Tf, all of which are present in FBS (Table 1). Optimal concentrations of both drug and cytokine are required to give rise to an optimal number of mature cells. Concentrations of drug and cytokines below the optimal level allow for some proliferation, whereas drug concentrations above that level cause increased cell death, leading to loss of selectivity. These concentration-dependent changes are shown in schematic form in Fig. 1, which summarizes essentially superimposable experimental results accrued with diverse DNA-specific agents, including ara C, daunomycin and actinomycin D.

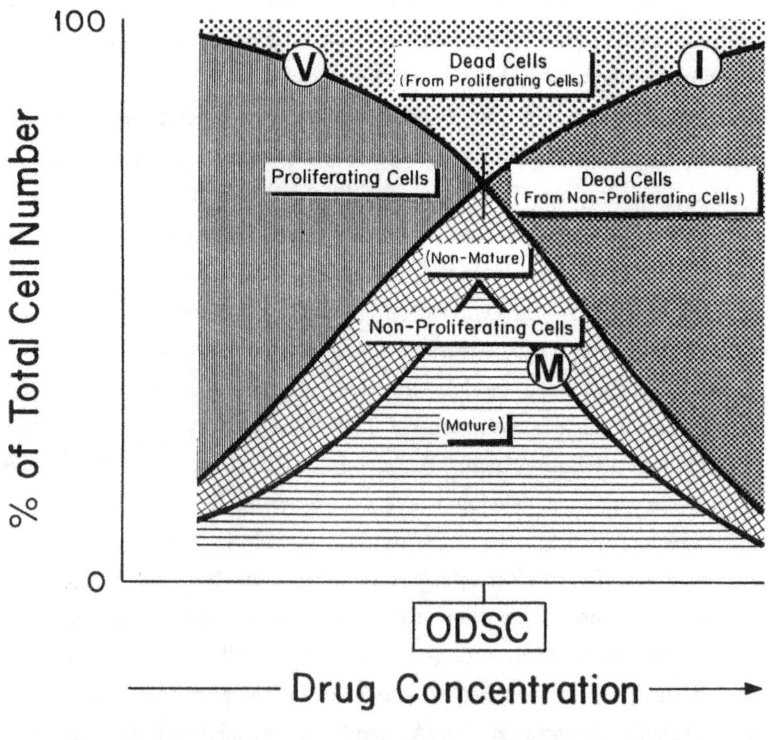

Figure 1

In the Figure, the viability curve V separates live from dead cells, the growth inhibiton curve I divides proliferating from non-proliferating cells, and the maturation curve M segregates the non-proliferating cell fraction into mature and non-mature cell compartments. While the mature cells have ceased to proliferate, cells in the non-mature segment can be committed to maturation or can be arrested reversibly or irreversibly in their growth, either naturally or by the presence of drug. The Figure shows that with increasing drug concentration the proliferating cell fraction decreases in size, the majority of its cells entering the non-proliferating cell compartment. Only a small fraction of proliferating cells die. A concentration of drug is reached at which the mature cell fraction (M) reaches its maximum. That concentration is defined as the optimal differentiation-sensitizing concentration (ODSC). Whereas the ODSC is a function of the drug, the size of the mature cell fraction that emerges at the ODSC is determined by the FBS or the cytokine concentration. At the ODSC, the proliferating cell fraction has ceased to exist, and since neither the mature cell segment nor the proliferation-derived dead cell fraction can contribute to the expansion of the cell population, it is only the reversibly arrested cells within the non-mature cell fraction that have the potential for renewal. Prolonged incubation of that fraction in the absence of drug did not result in measurable growth, indicating that regrowth, too, is optimally inhibited at the ODSC. At drug concentrations higher than the ODSC, non-proliferating cells are also killed, and they enter the dead cell compartment specified as being derived from non-proliferating cells.

What this scheme shows is that optimal inhibition of tumor cell growth can be achieved through induction of differentiation at drug concentrations that are only limitedly cytotoxic. If tumor cell populations are to be eliminated through cell kill, concentrations of drug higher than the ODSC are required, and these are less selective for the tumor cells. Thus, the fact that tumor cells are arrested at an immature stage of development provides an opportunity for selective therapeutic exploitation. Because initiation of maturation depends not only on drug but also on the presence of differentiation-inducing cytokines, both of these components need to be considered when treatments are formulated.

The mechanism by which DNA-specific agents stimulate differentiation-induction is related to the ultimate effect they exert on transcription. Thus, at optimal differentiation-sensitizing concentrations, antinomycin D caused a 38% decrease in total RNA, but an 87% reduction in rapidly turning-over mRNA [3]. The latter fraction comprises proliferation-related messages, including those specified by oncogenes such as myb. Likely a result of that decrease, the cells become more sensitive to the signal emitted by the low concentrations of differentiation-inducing cytokines present in the serum, to which - in the absence of drug - they are refractory. Sensitization to maturation signals as a mechanism of drug action is supported by the fact that the agents need to be applied prior to or simultaneously with the cytokines. If the cytokines are applied first, followed - after washing - by the drugs, differentiation is not induced. That it is, in fact, the cytokines TNF-α and TGF-β that participate in initiating the drug-sensitized differentiation process in ML-1 cells is shown by the observation that antibodies to these cytokines prevent the emergence of the mature cell fraction.

RNA- and protein-targeted agents, such as dichlorobenzimidazole riboside and cycloheximide, while inhibiting cell growth do not induce differentiation, supporting the notion that it is the inhibition of proliferation-regulating transcription that enhances the cytokine-mediated expression of stable, differentiation-related messages.

It may well be that the clinically established antitumor selectivity of DNA-specific agents relates to their ability to induce tumor cell maturation, possibly in conjunction with cytoreduction achieved through judiciously designed protocols.

Acknowledgments

The studies described were aided by Grant CA-36241 from the National Cancer Institute, DHHS.

REFERENCES:

1. Adesnik M, Snitkin H (1978) Induction of erythroid differentiation in Friend leukemia cells by bromodeoxyuridine. J Cell Physiol 95:307-318
2. Bodner AJ, Ting RC, Gallo RC (1981) Induction of differentiation of human promyeloctic leukemia cells (HL-60) by nucleosides and methotrexate. J Natl Cancer Inst 67:1025-30
3. Craig RW, Bloch A (1984) Early decline in c-myb oncogene expression in the differentiation of human myeloblastic leukemia (ML-1) cells induced with 12-0-tetradecanoylphorol-13-acetate. Cancer Res 44:442-446
4. Craig RW, Frankfurt OS, Sakagami H, Takeda K, Bloch A (1984) Macromolecular and cell cycle effects of different classes of agents inducing the maturation of human myeloblastic leukemia (ML-1) cells. Cancer Res 44:2421-2429
5. Denstman S, Hromchak RA, Guan X-P, Bloch A (1991) Identification of transferrin as a progression factor for ML-1 human myeloblastic leukemia cell differentiation. J Biol Chem 266:14873-14876
6. Friedman SJ, Skehan P (1979) Morphological differentiation of human choriocarcinoma cells induced by methotrexate. Cancer Res 39:1960-1967.
7. Fujii Y, Takuma T, Bloch A. (1990) A regulatory role for tumor necrosis factor (TNF) in ML-1 human myeloblastic leukemia cell maturation. Leukemia Res 14:941-947
8. Griffin J, Munro D, Major P, Kufe D (1982) Induction of differentiation of human myeloid leukemia cells by inhibitors of DNA synthesis. Exp Hemat 10:774-781
9. Guan X-P, Fujii Y, Hromchak RA, Bloch A (1989) Synergistic induction of ML-1 human myeloblastic leukemia cell differentiation by combinations of 12-0-tetradecanoylphorbol 13-acetate (TPA) with transforming growth factor (TGF-β) or with human tumor necrosis factor (TNF-α). Proc Am Assoc Cancer Res 30:175
10. Guan X-P, Hromchak RA, Takuma T, Denstman S, Bloch A (1991) Tumor necrosis factor-α, transforming growth factor-β and tetradecanoylphorbol acetate act as competence factors for ML-1 human myeloblastic leukemia cell differentiation. Cancer Communications 3:11-14
11. Gusella JF, Houseman D (1976) Induction of erythroid differentiation in vitro by purines and purine analogues. Cell 8:263-269
12. Honma Y, Honma C, Bloch A. (1986) Mechanism of interaction between antineoplastic agents and natural differentiation factors in the induction of leukemic cell maturation. Cancer Res 46:6311-6315
13. Hozumi M, Honma V, Tomida M, et al. (1979) Induction of differentiation of myeloid leukemia cells with various chemicals. Acta Hematol 42:941-952
14. Ishikura H, Hori K, Bloch A (1989) Differential biologic effects resulting from bimodal binding of recombinant human tumor necrosis factor to myeloid leukemia cells. Blood 73:419-424
15. Leder A, Orkin S, Leder P (1975) Differentiation of erythroluekemia cells in the presence of inhibitors of DNA synthesis. Science 190:893-894
16. Lotem J, Sachs L (1980) Potential pre-screening for therapeutic agents that induce differentiation in human myeloid leukemia cells. Intl J Cancer 25:561-564
17. Minowada J (1982) Immunology of leukemic cells. In: F. Gunz and E. Henderson (eds.) Leukemia 119-139

18. Nakaya K, Chou S, Kaneko M, Nakamura Y (1991) Topoisomerase inhibitors have potent differentiation-inducing activity for human and mouse myeloid leukemia cells. Japn J Cancer Res 82:184-91
19. Nishimune V, Kume A, Ogiso Y, Matsushiro A (1983) Induction of teratoma cell differentiation. Effect of Inhibitors of DNA Synthesis. Exp Cell Res 146:439-444
20. Papac RJ, Brown AE, Schwartz EL, Sartorelli NC (1980) Differentiation of human promyelocytic leukemia cells in vitro by 6-thioguanine. Cancer Letter 10:33-38
21. Rappa G, Lorico A, Sartorelli AC (1990) Induction of the differentiation of WEHI-3B D+ monomyelocytic leukemia cells by inhibitors of topoisomerase II. Cancer Res 50:6723-30
22. Sakagami H, Unten S, Konno K (1981) Close association of mouse myeloid leukemia (M1) cell differentiation with cessation of DNA synthesis. Cell Struct Funt 6:147-157
23. Schubert D and Jacob F (1972) 5-Bromodeoxyuridine-induced differentiation of a rat glioma in vitro. Cancer Res 32:1422-1427
24. Schwartz EL, Ishiguro K, Sartorelli AC (1983) Induction of leukemia cell differentiation by chemotherapeutic agents. Adv Enz Reg 21:3-20
25. Takeda K, Minowada J, Bloch A (1982) Kinetics of appearance of differentiation in ML-1, a line of human myeloblastic leukemia cells, after treatment with a 12-0-tetradecanoyl-phorbol-13-acetate, dimethyl sulfoxide, or 1-β-D arabinofuranosyl-cytosine. Cancer Res 42:5152-5158
26. Takeda K, Minowada J, Bloch A (1983) The role of a drug-induced differentiation in the control of tumor growth. In: Proc 9th Intl Symp on the Biol Characterization of Human Tumor. Bologna 1981, Berlin Akademie Verlag, 275-281
27. Takeda K, Minowada J, Bloch A (1991) Differential ability of mitogen-stimulated human leukocyte-conditioned media to induce F_c receptors in human leukemia cells. Cell Immunol 79:288-297
28. Terada M, Epner E, Nudel U, et al. (1978) Induction of murine erythroleukemia cell differentiation by actinomycin D. Proc Natl Acad Sci USA 75:2795-2799.
29. Waxman S, Scher BM, Hellinger N, Scher W. (1990) Combination cytotoxic-differentiation therapy of mouse erythroleukemia cells with 5-fluorouracil and hexamethylene bisacetamide. Cancer Res 50:3878-87
30. Waxman S, Schreiber C (1983) In vitro induction of differentiation of bone marrow by chemotherapeutic agents in myeloproliferative disorders. Blood 62:156

ATRA THERAPY IN ACUTE PROMYELOCYTIC LEUKEMIA A MODEL FOR DIFFERENTIATION THERAPY

Laurent DEGOS
Hematology Department
SAINT LOUIS Hospital
1, avenue Claude Vellefaux
75010 PARIS - FRANCE

Leukemia are characterized by a clonal expansion of cells which are arrested at an early stage of maturation. The immature cells infiltrate the bone marrow and inhibit the normal hematopoiesis. The mitotic index of malignant cells is often similar or lower than the index of the normal cells indicating that the increased number of malignant cells in the bone marrow and in the blood is due to an accumulation more than to a proliferative effect. The accumulation is the consequence of a long survival of malignant cells probably due to an impairment of the cell death program related to the immaturity of the cells.

Acute promyelocytic leukemias represent 10 % of non acute lymphoblastic leukemia (acute myeloblastic leukemia M3 subtype of FAB nomenclature (Larson et al, 1984)), and is characterized by a translocation between the chromosome 17 and 15 (Bennet et al, 1976).

Leo Sachs demonstrated that the differentiation arrest is sometimes reversible when leukemic cell lines are treated in short term culture with agents inhibiting the proliferation or enhancing the differentiation (Sachs, 1978). We have studied various antimitotic at low concentration and we found that cytosine arabinoside is able to stimulate the differentation of cell lines but with a less efficacy to induce a maturation of fresh cells from patients.

Retinoic acids are able to differentiate malignant and normal tissues (Lotan, 1980). Theodore Breitman first reported the maturation of the granulocytic HL-60 cell line, and of leukemic cells from patients with APL. With Christine Chomienne we confirmed, in a series of bone marrow specimens from 60 patients with leukemia, that retinoic acids are able to differentiate specifically cells from cases of APL (Chomienne et al, 1990).

NATO ASI Series, Vol. H 75
Cancer Therapy
Edited by N. D'Alessandro, E. Mihich, L. Rausa,
H. Tapiero, and T. R. Tritton

We have compared the effects of various derivatives of vitamin A, 13-cis, all-trans and 4 OXO retinoic acid at different concentration 10^{-6} M, 10^{-7} M and 10 $^{-8}$M, and we found that at 10^{-7} M and at 10^{-8} M all trans retinoic acid gives better results than the other derivatives (Chomienne et al, 1990).

I - CLINICAL RESULTS

I.1. Clinical activity :

All-trans retinoic acid was not available in Western countries. For this reason we initiated a collaboration with Professor WANG Zhen Yi from Shanghaï n°2. He treated first newly diagnosed patients using various doses ranging from 10 to 100 mg/m2 and he reported 23 complete remissions (CR) among 24 patients (Huang et al, 1988). We then treated patients in first relapse of APL with the chinese drug and we obtained 19 complete remissions among 20 patients using ATRA at a fixed dose (45 mg/m2) during 3 months (Degos et al, 1990). Up to now, more than 1000 patients have been treated with ATRA. The mean of CR rate in newly diagnosed patients and in patients in first relapse is over 90 % (even in patients who resist to conventional therapy) (Fenaux et al, 1991). However we found a weaker activity in patients in second or subsequent relapses (Castaigne et al, 1990).

II.2. The activity of ATRA is a differentiation of malignant cells :

The complete remission is achieved without any feature of an aplastic phase. For instance in the first relapse patients (Degos et al, 1990) 75 % of them were hospitalized for less than a week, 70 % did not require any transfusion, sterile conditions or antibiotics. Only an oral treatment was prescribed at home.

Morphologic changes of malignant cells are documented by serial examinations of bone marrow (Castaigne et al, 1990). Maturing cells with apparent terminal differentiation progressively appeared. Moreover Auer rods were sometimes present in mature cells confirming the differentiation process. Simultaneously, normal cells reoccured in the bone marrow. The cytogenetical follow up disclosed a progressive disappearance of the abnormal clone and a normal karyotype at the time of the complete remission.

An intermediate population expressing both mature (CD16) and immature (CD33) markers during the 3rd and 4th week of treatment was found using serial studies of cell surface immunophenotyping (Warrell et al, 1991). In situ hybridization with a chromosome 17 probe confirmed the relationships between the clinical response and the maturation of the leukemic clone.

II.3 The hyperleucocytosis is the major adverse effect :

Minor adverse effects are generally reported which have been previously reported in RA treatment for skin diseases : dryness of the skin and mucosae, headache, transient bone marrow pain, increases of triglycerides and transaminases. These symptoms are easily treated by cream and eye drops, or by analgesics.

The hyperleucocytosis is the major adverse effect of retinoic acid treatment and seems to be specific to the APL treatment (Castaigne et al, 1990). A mild increased of white blood cells is found in almost all patients during the 3 first weeks of treatment. High hyperleucocytosis, not reported in chinese results, occurs in 40% of de novo patients (personnal data) and in 15 to 20 % first relapse patients (Degos et al, 1990). In a series of 22 patients, Castaigne et al reported 3 early deaths, all of the patients having more than 30 x 10^9/L WBC. In the subsequent treated patients among 5 patients with WBC exceeding 30 x 10^9/L, 3 patients died (personnal data). Hyperleucocytosis is not related to a resistance : the WBC are maturing cells, and the two patients who survived had a progressive normalization after the third week under ATRA treatment.

A specific "retinoic acid syndrome" (Frankel et al, 1992) occuring during this hyperleukocytic phase includes fever, pulmonary infiltrates, respiratory distress, kidney failure, and a progressive coma. Autopsy revealed an organ infiltration with myeloid cells, focal hemorrhages and thrombosis. Short course, high dose corticosteroid treatment promptly reverse these symptoms in four among five reported cases treated in the Memorial Sloan Kettering Institute (Frankel et al, 1992). It was previously proposed by the Memorial Sloan Kettering Institute to prescribe a leukapheresis but this method is no more applied.

I.4. The coagulation disorders are rapidly corrected

The major beneficial clinical effects are the rapid disappearence of the hemorrhagic syndrome and the absence of bone marrow hypoplasia.

The bleeding diathesis was attributed to a disseminated intravascular coagulation. However primary fibrinolysis may play a major role in bleeding, suggested by low level of fibrinolytic inhibitors and an extensive proteolysis. ATRA therapy corrects the fibrinolysis with a rapid normalization of fibrinogen level, while symptoms of DIC persist during several weeks. The bleeding diathesis is rapidly removed, but the procoagulant tendency persists during the first month of treatment. The ATRA therapy is able to distinguish these two disorders, the primary fibrinolysis and the DIC (Dombret et al, in press).

I.5. A resistance appears progressively during ATRA treatment

The progressive acquired resistance explains the shortness of the remission duration and a weaker activity when ATRA is used for subsequent relapses. The median duration of complete remission is 5 months in our experience if ATRA is given continuously. The resistance to ATRA seems not to be associated with additional genetic modifications. The resistance may in part be explained by a marked decrease in plasma drug concentration (Warrell et al, 1991). Mechanisms for accelerated clearance include induction of catabolic cytochrome P450 oxydative enzymes, along with increased expression of cellular retinoic acid binding protein (CRABP) that sequestres ATRA in the cytoplasm and enhances its catabolism.

We have found a progressive increase of the CRABP in the cytosol of all normal and abnormal myeloid cells (Cornic et al, 1992). The failure to maintain a clinical response could be due to the inability to sustain a differentiating concentration in the nucleus, where the functional receptors are located.

I.6. Treatment strategy of APL

In the French pilot study when 26 patients received ATRA, 25 obtained a complete remission (95%) but 11 had an early chemotherapy due to the hyperleucocytosis (40%). One patient did not received a consolidation and relapsed rapidly. Among the 25 remaining patients in complete remission, 20 remained alive and in complete remission 2 years later (80 %). Three patients had a relapse of the disease (8, 11, 15th months) and two died in complete remission (sepsis for consolidation, and conditioning treatment for an allograft). The actuarial disease free survival in this series is 83% and the event free survival is 70% at two years (Fenaux et al, in press).

Our proposal for the strategy in the treatment of acute promyelocytic leukemia combines ATRA as a differentiating agent in order to obtain a complete remission avoiding the early deaths and a chemotherapy as a consolidation therapy in order to maintain the complete remission. The chemotherapy could be used earlier if an hyperleucocytosis occurs in order to prevent the retinoic acid syndrome. The bleeding diathesis seems to be rapidly recovered by the ATRA treatment. However it remains a tendency of procoagulant activity with the risks of thrombosis which could be prevented by low molecular weight heparin.

An European trial for de novo patients with APL started in April 1991 compares the conventional chemotherapy to ATRA before chemotherapy with the same guidelines as the pilot study in order to avoid hyperleucocytosis (Pierre

Fenaux, Lille, France). The consolidation includes 2 courses of Daunorubicin and Cytosine Arabinoside. A similar study is initiated in USA, Canada and Australia launched in April 1992, with a second randomization after the consolidation phase, testing a maintenance (or not) with ATRA (Peter Wernick NY USA).

Our three major recommandations for the treatment of APL are the following :

1. Use ATRA first in order to avoid early deaths and then a chemotherapy is added for "long survival".
2. Prevent the hyperleucocytosis (chemotherapy according to the WBC count) and/or treat the retinoic acid syndrome (corticosteroids).
3. Follow the procoagulant activity (DIC) and thrombosis which are treated by heparin if necessary.

Some questions are not yet solved : How long ATRA must be given ? How to combine ATRA and chemotherapy (sequentially or simultaneously) ? Is a maintenance therapy necessary ? Which kind of maintenance therapy ?

II - MOLECULAR BIOLOGY

Acute promyelocytic leukemia is consistently associated with a balanced and reciprocal translocation between the long arms of chromosome 15 and 17. The retinoic acid receptor alpha and many other genes as myeloperoxydase, G-CSF, ErbA, were shown to be mapped close to the breakpoint in 17q21. The breakpoint location in APL and the clinical activity of ATRA prompted us to investigate the gene structure of the retinoic acid receptor alpha (RARa) using a probe kindly provided by Pierre Chambon.

Christine Chomienne et al (1990) first recognized, by the presence of an additional band in Northern blots, the specific rearrangement of messenger RNA of RARa in APL, not present in the other types of leukemia or in normal granulocytes.

Michel Lanotte et al (1991) made a cell line from a patient having the (15;17) translocation and Hughes de Thé et al (1990) cloned and sequenced the translocation site. They demonstrated that the RAR is rearranged at the second intron and fused to a not known gene on chromosome 15 that we called myl and renamed PML for promyelocytes. Synthesis of two reciprocal fusion transcripts

(PML RARa found in all cases of APL and RARa/PML found in two third of cases) are the consequences of this fusion.

Sylvie Castaigne et al (1992) developped an RT-PCR technique, using two specific primers, one for the RARa, the other for PML and demonstrated the presence of the hybrid PML-RAR transcript in all cases of APL. The presence of an amplified small sequence of DNA gives the evidence of the specific rearrangement. This method is actually used for the diagnosis of APL and for the follow up of the minimal residual disease.

Retinoic acid receptors are nuclear receptors which bind to specific DNA sequences in the promotor region of target genes. They are able to modulate the transcription of target genes when the ligand (RA) is present. The fusion protein PML-RAR on one side contains a large part of the RARa with the domain involved in the retinoic acid binding and the domain that binds to the DNA sequences. However the first domain probably involved in the transactivating activity is deleted. On the PML side, the breakpoints on chromosome 15 are clustered in two different introns and one exon leading to distinct fusion proteins grouped in long and short PML-RAR. Both contains a zinc finger DNA binding sequences and a leucine zipper domain.The two molecules RARa and PML being DNA binding proteins seem to be involved in the regulation of transcription of specific sets of genes. Furthermore both of them are characterized by the formation of dimers : RAR with RAR (homodimers) or with RXR (heterodimers) and PML with other genes due to the leucine zipper sequence similar to the jun-fos model of AP1. A PML-RAR fusion product should alter the normal transactivating activity of PML and RAR.

We approached the transforming potential of the fusion gene by co transfection assays demonstrating that PML-RARa has not functioned as a normal RAR (de Thé et al, 1991). The cells were first transfected with retinoic acid sensitive reporter made by responsive elements, which are the DNA sequences used by the RAR for the binding, linked to a signal sequence (TK-luciferase). In presence of retinoic acid, endogenous receptors are able to induce a positive signal. A subsequent transfection with the normal receptor gene increases the signal while a transfection with PML-RAR extinguishes the signal. The fused protein is able to displace the normal one and alters the transactivating activity on target genes.

Farzin Farzeneh and Christine Chomienne (not published) recently used liposome method for the gene transfection and demonstrated that in a myeloid cell line the fused gene PML-RAR inhibited the induced maturation by RA but not by other agents as DMSO or TPA.

This inhibition of differentiation could explain the arrest of maturation in the leukemic cells. All these experiments explored the role of RA and of RAR. Due to the ignorance of responsive elements and of target genes for PML gene nothing is known on the role of the impairment of PML in the pathogenesis of the disease.

Why retinoic acids are able to revert the malignancy ?

We approached this question by two ways.

Firstly, we found that the inhibition of PML-RAR in transfection assays in presence of retinoic acid sensitive reporter is overpassed when retinoic acid is added in high concentration, compatible with the pharmacological doses given to the patients (Chomienne et al, 1991).

Secondly, we noted that in APL cells, PML/RARa appears to be considerably more abundant than the normal receptor encoded by the intact chromosome 17. In presence of retinoic acid, in short term culture, we found a rapid increase of the expression of the normal receptor (Chomienne et al, in press). An hypothesis could be a reversion of the inhibition made by the aberrant protein in presence of large amount of the normal protein, sufficently abundant to displace the abnormal one on the DNA responsive elements. Therefore, high concentration of retinoic acid restores a normal transactivation on target genes by the induction of functional activity of the abnormal PML-RAR hybrid molecule and/or a functional level of the normal transcript of RAR gene. Leukemic cells from patients better survive in short term culture in presence of retinoic acid. They are engaged in the differentiation process. We also found that they are also engaged in their cell death program, looking at the expression of Bcl-2 protein. Bcl-2 is strongly expressed in all leukemic cells. After few days of culture in presence of RA the intensity of the expression is dramatically decreased. Bcl-2 is known to be a gene involved in the cell death program which is engaged when the gene expression is switch off (Chomienne et al, in press). Retinoic acid removing the target gene functions could restore the control of the cell death program.

III - CONCLUSION

ATRA treatment induces a differentiation of malignant cells in all cases of acute promyelocytic leukemia. Patients achieved complete remissions without aplastic phase and with a rapid recover of bleeding diathesis. A strategy of treatment combines ATRA first and a consolidation by chemotherapy in order to avoid early deaths and to maintain a long term survival. Retinoic acid receptor alpha is rearranged by the translocation 15;17. A fusion transcript PML/RAR is made and the hybrid protein alters the transactivating activities of normal retinoic acid receptor. Moreover the hybrid protein impairs the granulocytic maturation induced by retinoic acids. High concentration of retinoic acid restores a normal function of the abnormal protein and induces a high level of expression of the normal gene. The normal activity of target genes engages the cell to a maturation process, to a terminal differentiation and to a normal regulation of cell death program. Thus, the malignancy is not treated by cytotoxic events, but by the restoration of the normal cell biology. ATRA therapy in APL represents the first model of differentiation therapy in human malignancies.

BIBLIOGRAPHY

Bennet JM, Catowsky D, Daniel MT, Flandrin G, Galton DAG, Gralnick HR, Sultan C (1976) Proposals for the classification of the acute leukemia. Br J Haematol 33:451-458

Castaigne S, Chomienne C, Daniel MT, Ballerini P, Berger R, Fenaux P, Degos L (1990) All trans retinoic acid as a differentiation therapy for acute promyelocytic leukemia. I. Clinical results. Blood 76: 1704-1709

Castaigne S, Balitrand N, de Thé H, Dejean A, Degos L, Chomienne C (1992) A PML/RAR alpha fusion transcript is constantly detected by RNA-based polymerase chain reaction in acute promyelocytic leukemia. Blood 79: 3110-3115.

Chomienne C, Ballerini P, Balitrand N, Daniel MT, Fenaux P, Castaigne S, Degos L (1990) All trans retinoic acid in acute promyelocytic leukemia. II. In vitro studies : structure function relationship. Blood 76: 1710-1717

Chomienne C, Ballerini P, Balitrand N, Huang ME, Krawice I, Castaigne S, Fenaux P, Tiollais P, Dejean A, Degos L, de The H (1990) The retinoic acid receptor alpha gene is rearranged in retinoic acid sensitive promyelocytic leukemia. Leukemia 4: 802-807

Chomienne C, Balitrand N, Ballerini P, Castaigne S, de Thé H, Degos L (1991) All trans retinoic acid modulates the retinoic receptor alpha in promyelocytic cells. J Clin Invest 88: 2150-2154

Chomienne C, Barbey S, Balitrand N, Degos L, Sachs L (1992) Regulation of Bcl-2 and cell death by all-trans retinoic acid in acute promyelocytic leukemic cells. AACR Abstract (in press)

Cornic M, Delva L, Guidez F, Balitrand N, Degos L, Chomienne C (1992) Induction of retinoic acid binding protein in normal and malignant human myeloid cells by retinoic acid in AML3 patients. Cancer Res 52: 3329-3334.

Degos L, Chomienne C, Daniel MT, Berger R, Dombret H, Fenaux P, Castaigne S (1990) Treatment of first relapse in acute promyelocytic leukemia with all trans retinoic acid. Lancet 336: 1440-1441

De Thé H, Chomienne C, Lanotte M, Degos L, Dejean A (1990) The t(15;17) translocation of acute promyelocytic leukemia fuses the retinoic acid receptor alpha gene to a novel transcribed locus. Nature 347: 558-561

De Thé H, Lavau C, Marchio A, Chomienne C, Degos L, Dejean A (1991) The PLM-RARα fusion mRNA generated by the t(15;17) translocation in acute promyelocytic leukemia encodes a functionnaly altered RAR. Cell 66: 675-684

Dombret H, Scrobohacci ML, Gorrha P, Zini JM, Daniel MT, Castaigne S, Degos L (1992) Coagulation disorder associated with acute promyelocytic leukemia : corrective effect of all trans retinoic acid treatment. Leukemia (in press)

Fenaux P, Degos L (1991) Treatment of acute promyelocytic leukemia with all trans retinoic acid. Leuk Res 15: 655-657

Fenaux P, Castaigne S, Dombret H, Chomienne C, Duarte M, Archimbaud E, Lamy T, Tibeghien P, Tilly H, Dufour P., Cransac M, Guerci A, Sadoun A, Degos L (1991) All-trans retinoic acid in newly diagnosed acute promyelocytic leukemia: a pilot study. Blood (ASH Abstract) Abst n° 305 (plenary session) p. 79a, 1991;, and Blood 1992 (in press)

Frankel S., Eardley A., Lauwers G., Weiss M., Warrell RP (1992) The retinoic acid syndrome in acute promyelocytic leukemia. Ann Int Med 117: 292-296.

Huang ME, Ye YC, Chai JR, Lu JX, Zhoa L, Gu LJ, Wang ZY (1988) Use of all trans retinoic acid in the treatment of acute promyelocytic leukemia. Blood 72: 567-572

Lanotte M, Martin Thouvenin V, Najman S, Ballerini P, Valensi F, Berger R (1991) NB4, a maturation inducible cell line with t(15;17) marker from a human acute promyelocytic leukemia M3. Blood 77: 1080-1086

Larson RA, Kondo K, Vardiman JW, Butler AE, Golomb HM, Rowley JD (1984) Evidence for a 15;17 translocation in every patients with acute promyelocytic leukemia. Am J Med 76:827-841

Lotan R (1980) Effect of vitamin A and its analogs (retinoids) on normal and neoplastic cells. Biochem Biophys Acta 605: 33

Sachs L (1978) Control of normal differentiation and the phenotypic reversion of malignancy in myeloid leukemia cells. Nature 274: 535

Warrell RP, Frankel SR, Miller WH, Scheinberg DA, Itri LM, HIttelman WN, Vyas R, Andreeff M, Tafuri A, JAkubowski A, Gobrilove J, Gordon MS, Dimitrovsky E (1991) Differentiation therapy of acute promyelocytic leukemia with tetrinoin (All trans retinoic acid). New Engl J Med 324: 1385-1393

Warrell RP, Muindi J, Frankel SR, Jakubowski A, Dmitrovsky E, Miller WH (1991) Continuous treatment with all trans retinoic acid progressively decreases plasma drug concentrations : implications for relapse and resistance in acute promyelocytic leukemia. Blood (ASH abstract) abst n° 1063, p. 268a

HEMIN IS TRANSPORTED IN HUMAN LEUKEMIA K562 CELLS AND INTERACTS WITH DNA SEQUENCES.

Asterios S. Tsiftsoglou[1*], Athina I. Tsamadou[1], Stephen H. Robinson[2] and Willie Wong[2]
[1]Laboratory of Pharmacology, Department of Pharmaceutical Sciences, Aristotle University of Thessaloniki, Thessaloniki, 54006, Greece and [2]Department of Medicine, Beth Israel Hospital and Harvard Medical School, Boston, MA. 02215

INTRODUCTION

Heme (ferroprotoporphyrin IX), a natural agent, serves as a prosthetic group in various hemoproteins like hemoglobin, myoglobin, cytochromes, catalase and others involved in oxygen transport, cellular respiration, ATP production and drug metabolism (Stryer, 1988). Hemin, the oxidised form of heme, has been shown to activate gene expression and promote differentiation in a variety of cell types including mouse 3T3 cells (Chen and London, 1981), neuroblastoma (Ishii et al, 1978), erythroleukemia (MEL, K562) cells (Ross et al, 1976; Rutherford et al, 1979a; Tsiftsoglou and Robinson, 1985; Tsiftsoglou et al, 1991) as well as normal hematopoietic cells like CFU-E, CFU-GM and BFU-E (Monette et al, 1982) by interacting with cellular components at various levels (Sassa et al, 1988).

Among the biological effects, the induction of hemoglobin in MEL and K562 leukemia cells by hemin has attracted a lot of attention in recent years (Ross et al, 1976; Rutherford et al, 1979; Tsiftsoglou and Robinson, 1985; Charney and Maniatis, 1983; Tsiftsoglou et al, 1989). These observations offered suitable model systems to uncover the molecular mechanism(s) of activation of globin gene expression by hemin and study hemoglobin switching (Rutherford et al, 1979a; Charney and Maniatis, 1983; Tsiftsoglou et al, 1989). Hemin stimulates the production of both embryonic (Gower I,

*Correspondence must be addressed to Prof. Asterios S. Tsiftsoglou

NATO ASI Series, Vol. H 75
Cancer Therapy
Edited by N. D'Alessandro, E. Mihich, L. Rausa, H. Tapiero, and T. R. Tritton

Portland) and fetal hemoglobins (HbF, Barts) in substantial quantities without inducing synthesis of adult type hemoglobins (HbA_1, HbA_2) and promoting terminal maturation (Rutherford et al, 1979a; Tsiftsoglou et al, 1989; Benz et al, 1980). Although hemin does not initiate commitment in MEL cells (Gusella et al 1980), Rowely et al (1985) have claimed that hemin promotes commitment of K562 cells to terminal erythroid maturation to a low degree. By acting cooperatively with aclacinomycin, a member of the anthracycline class agents, hemin promotes terminal erythroid maturation and induces production of immature β-globin RNA transcripts in K562 cells due to incomplete termination of transcription of β-like globin DNA sequences (Tsiftsoglou et al, 1989; Tsiftsoglou et al, 1991).

The precise mechanism(s) by which hemin enters K562 cells and activates transcription of globin genes are not well understood. Studies carried out with the use of radiolabeled hemin in murine erythroleukemia cells (MEL) by Galbraith et al (1985) and most recently in K562 cells by us (Tsamadou et al, 1992) indicated that hemin is taken up by both of these cells via a carrier-mediated process. The hemin-induced production of hemoglobin synthesis in K562 cells is reversible (Dean et al, 1981) and appears to result from direct or indirect action of hemin on transcription of globin DNA sequences. In fact several lines of investigation thus far have suggested that hemin may exert some of its biological effects by interacting at the nuclear level. Hemin does not control total protein synthesis and globin synthesis via a hemin-controlled repressor (London et al, 1981; Rutherford et al, 1979b) but regulates transcription of the iso-1-cytochrome c (CYC1) gene in yeast via a protein that binds to an upstream activation site (UAS) (Guarente et al, 1983). In this study, we have attempted to determine: (a) the kinetics of hemin-uptake in K562 cells and its accumulation into cytosol and nucleus; (b) the interaction of hemin with nuclear components, and (c) the effect of hemin on the interactions of globin DNA sequences with cytoplasmic and nuclear trans-acting factors.

RESULTS

[^{14}C]-hemin is transported into K562 cells, accumulates into the nucleus and interacts with DNA.

As shown in Fig. 1 and as indicated earlier (Tsamadou et al, 1992), the use of radiolabeled hemin permitted us to demonstrate that [^{14}C]-hemin enters the K562 cells and accumulates intracellularly by reaching a maximum level within 60 min. Quite interestingly, we observed that a substantial portion of [^{14}C]-hemin was rapidly associated with the cells and accumulated into the cytosol thereafter. This suggests that a portion of hemin is transported into K562 rapidly. The major portion of intracellular hemin remained into cytosol and substantial quantities of [^{14}C]-hemin were detected

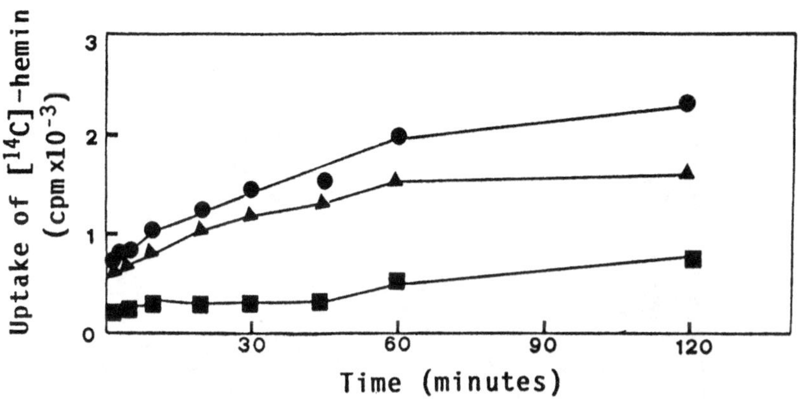

Figure 1. *Uptake of [^{14}C]-labelled hemin in cytosol nucleus and intact K562 cells.* [^{14}C]-labelled heme was prepared as previously (Tsamadou et al, 1992) and used for uptake and equilibrium membrane dialysis studies. Human leukemia K562 cells originally developed by Lozzio and Lozzio (1975) were used throughout this study as described by Tsiftsoglou et al (1989). Exponentially growing K562 cells were labelled with [^{14}C]-labelled hemin ($2.5x10^5$ cpm/ml). At various times following incubation, duplicate aliquots of cell suspension (0.4 ml) were removed, layered over an ice cold 24% w/v sucrose Na^+ Ringer solution and centrifuged at 13.000xg for 3 min. Cell pellets were lysed with 200 µl 1% SDS solution and counted for radioactivity. To determine the intracellular distribution of transported hemin, cells harvested by centrifugation at 13.000xg throughout the sucrose solution were subsequently fractionated into cytoplasm and nuclei with the use of lysis buffer (0.14 M NaCl, 1.5 mM $MgCl_2$, 10 mM Tris-HCl pH 8.6, 1% NP-40). Radioactivity was counted in each fraction.

[^{14}C]-hemin in intact cells (●—●), cytosol (▲—▲) and nucleus (■—■).

into the nucleus later on (after 45-60 min). These studies indicate that hemin is transported into the K562 cells easily, accumulates into cytosol first and then moves into the nucleus.

To further verify that [^{14}C]-hemin interacts with nuclear components, we prepared nuclei from control and hemin-treated K562 cells and incubated them with increasing concentrations of [^{14}C]-hemin. As shown in Fig. 2, the amount of [^{14}C]-hemin associated with isolated nuclei increased as a function of the concentration of [^{14}C]-hemin added. Less [^{14}C]-hemin was associated with the same number of nuclei prepared from hemin-treated cells as compared to that associated with nuclei from control cells. This difference may be due to preoccupation of possible hemin binding sites by exogenously non-labeled hemin added in the culture medium during incubation of K562 cells with hemin.

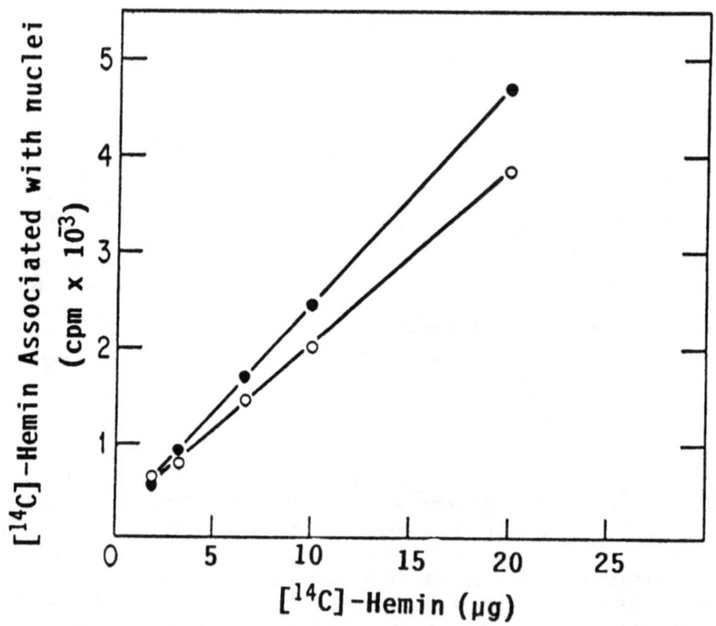

Figure 2. *Association of [^{14}C]-hemin with intact nuclei prepared from K562 cells.* Constant number of nuclei (1x10^{6}) prepared from control K562 cells and cells treated with hemin (30 μM, purchased from Eastman Kodak, NY) (see Tsiftsoglou et al, 1981) were incubated with varying concentrations of [^{14}C]-hemin (40.000 cpm/10μg) at 37°C for 30 min. The amount of [^{14}C]-hemin associated with nuclei was measured as follows: by the end of the incubation period, the mixture was centrifuged at 12.000 x g, washed with phosphate buffer saline pH 7.0 (PBS), resuspended in 0.1 ml of 1 M NaOH, boiled for 1 min, neutralized with 0.1 ml 1 M HCl and counted for radioactivity using Aquasol a liquid scintillation cocktail.

[^{14}C]-Hemin associated with nuclei prepared from control (●——●) and hemin-treated cells (○——○).

Association of [^{14}C]-hemin with purified nuclei could be due either to direct interaction of hemin with DNA or nuclear proteins or both simultaneously. To demonstrate whether [^{14}C]-hemin indeed interacts with DNA from

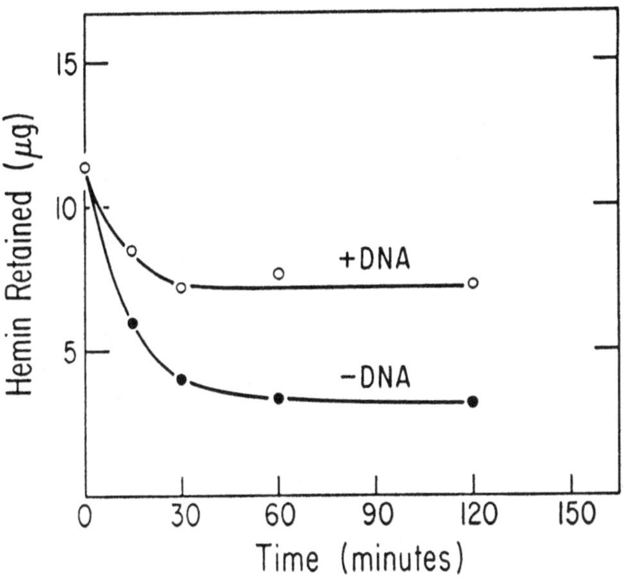

Figure 3. *Membrane equilibrium dialysis study of hemin-DNA interactions.* [^{14}C]-hemin (40.000 cpm/10μg) was incubated with 40 μg of sheared DNA from K562 cell in a final volume of 500 μl 10 mM Tris-HCl pH 7.4/1 mM EDTA for 30 min at 37°C. The mixture was then transferred into a dialysis bag (mol. weight cut off: 12.000, Spectra Fisher Scientific Co.) and the amount of [^{14}C]-hemin retained inside was determined as a function of time of dialysis.

Hemin retained in the absence (●——●) and presence of K562 cellular DNA (○——○).

K562 cells, DNA isolated from control untreated cells and sheared into relatively low mol. weight (-500 kb) fragments was incubated in [^{14}C]-hemin at 37°C. When the mixture was membrane dialyzed and the kinetics of [^{14}C]-hemin efflux was measured, we observed that less [^{14}C]-hemin was retained within the membrane dialysis bag in the absence of DNA (Fig 3). These results indicate that DNA retarded the efflux of [^{14}C]-hemin. So higher levels of [^{14}C]-hemin retained presumably due to direct interaction between these two agents. Subsequent Sephadex G-150 analysis of the mixture confirmed the presence of a [^{14}C]-hemin-DNA complex indicating that [^{14}C]-hemin interacts directly with DNA like with proteins prepared from K562 cells as recently indicated (Tsamadou et al, submitted for publication). A third piece of evidence shown that hemin and DNA interact each other stems from preliminary studies showing that a portion of sheared DNA attached on and eluted from a heme-agarose affinity column in contrast to same DNA loaded on a "plain" agarose column (Tsiftsoglou, unpublished observations).

Hemin prevents cytoplasmic and nuclear trans-acting fractions of K562 cells to interact with discrete regions of β-globin DNA sequences.

Earlier studies reviewed by Orkin (1990) have established that the β-like globin genes in humans are organised in a family of genes located on chromosome 11 as indicated in Fig. 4. All β-like globin genes (embryonic, fetal, and adult) are linked to each other and are under the influence of

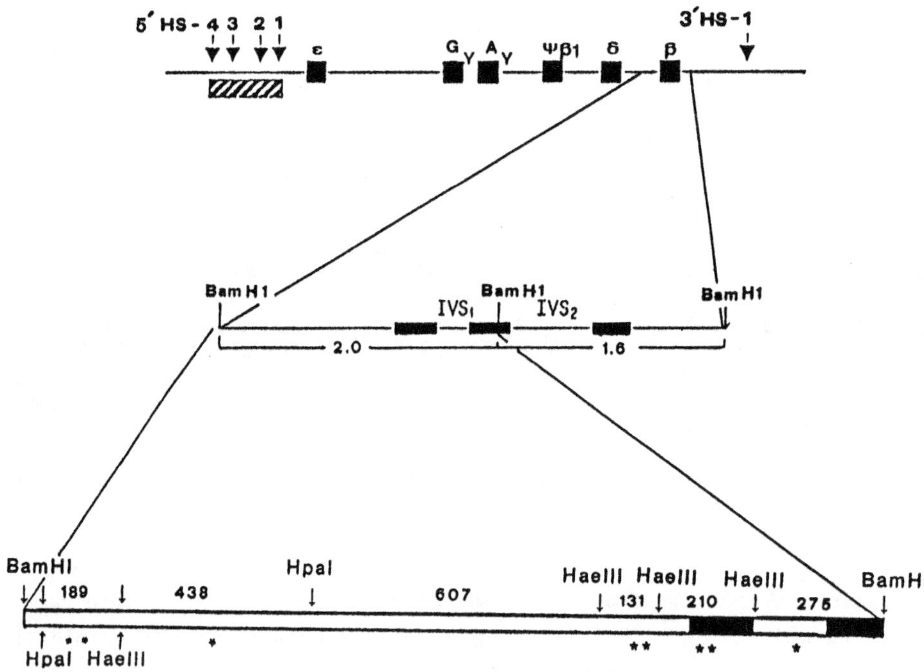

Figure 4. *Diagrammatic representation of the organisation of human β-globin gene family on chromosome 11 and restriction map of the 3.6 kb BamH1-BamH1 fragment of β-globin DNA used for this study (Treisman et al, 1983; Lown et al, 1980).* The asterisks indicates possible sites of interaction with cytoplasmic and nuclear factors. The β-like globin genes are organised in family of genes on chromosome 11 and are under the influence of local control region (LCR) region via by trans-acting activators like GATA-1 proteins and as yet unknown repressors (Orkin, 1990).

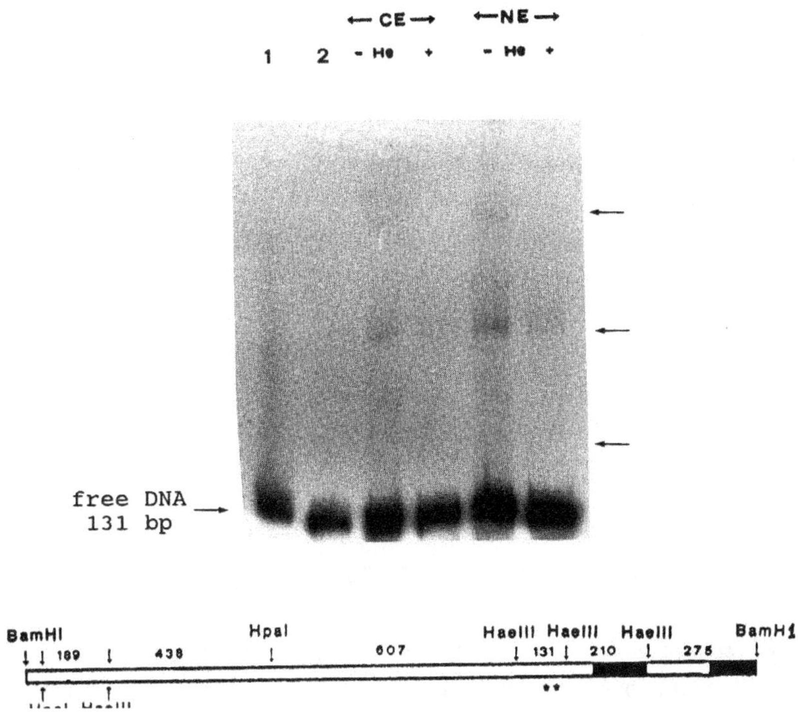

He: Hemin

CE: Cellular extracts
NE: Nuclear extracts

1: DNA
2: DNA/poly (dI-dC)/hemin

Figure 5. *Effects of hemin on the interaction of cytoplasmic and nuclear factors with the 131bp (HaeIII-HaeIII) region of the promoter of human β-globin DNA sequences.* Soluble cytoplasmic extracts were prepared from cultured K562 cells which were lysed with a solution containing 10 mM Tris-HCl pH 7.4, 1 mM $CaCl_2$, 7% sucrose in 0.9% NaCl, 1 mM PMSF, and 1% NP-40 and centrifuged 16.000 g for 30 min at 4°C to remove nuclei and to yield the post mitochondrial fraction. Nuclear extracts were prepared from $2x10^8$ cells essentially as described by Dignam et al (1983). Different size fragments of DNA excised from the 3.6 kb β-globin DNA (BamH1 cut) by further digestion with restriction enzymes were 5'-end labelled as described by Maniatis and Sambrook (1983). One of them, the [^{32}P]-labelled 131 bp DNA fragment was incubated separately with cytoplasmic and nuclear extracts (5-10 μg protein) from K562 cells in the presence or absence of hemin (30 μM). The mixture was analysed electrophoretically according to Garner and Revzin (1981). Finally, gels were transferred to Whatman 3MM, dried up and autoradiographed. The arrows indicate formation of complexes.

the main local control region (LCR), a locus enriched in several hypersensitive sites (HSs). Although each β-like globin gene has its own promoter, overall the transcription of each globin gene of the β-family is under the influence of LCR which is regulated via trans-acting factors like GATA proteins and repressors (silencing activities). Trans-acting factors which interact with globin genes recognise certain DNA-binding domains (motifs). A 3.6 kb BamH1 fragment containing the exon I, IVS1, exon II, IVS2, exon III and flanking sequences of β-globin gene was used to generate several smaller fragments with the use of restriction enzymes as shown in Fig 4. It is reminded that hemin activates but not terminates the transcription of β-globin DNA sequences in K562 cells (Tsiftsoglou et al, 1989). Therefore, it was in our interest to determine whether K562 cells contain cytoplasmic as well as nuclear trans-acting factors that interact with 5′- or 3′-end [^{32}P]-labelled fragments of β-globin DNA sequences. As shown in Fig. 5 incubation of the HaeIII 131 bp fragment obtained from the promoter region of the β-globin DNA sequence with soluble cytoplasmic as well as nuclear extracts, indicated that there may be at least three trans-acting factors forming complexes with this region of globin DNA sequences. Complex formation, however, occurred only in the absence of hemin. Addition of hemin abrogated formation of such complexes almost in every case examined. This suggests that hemin prevents the interaction of cytoplasmic (CE) as well as nuclear (NE) transacting factors with the 131 bp fragment as illustrated in Fig. 5. According to gene sequence data published elsewhere (Lawn et al, 1981), the 131 bp region contains the consensus sequences CACCC and TATCT. Similar DNA gel retardation assays carried out with other fragments of β-globin DNA sequences shown in Fig. 5 indicated that K562 cells contain several trans-acting factors which interact with sequences located within the fragments of 189, 210, 275 and 438 bp as shown by the asterisks (see Fig. 4), but not in that of 607 bp (Tsiftsoglou et al, in preparation).

DISCUSSION

As indicated by several studies over the years, heme as well as hemin exert pleiotropic effects on a number of cell types. Therefore, hemin may well be considered a natural regulator of growth and differentiation of various cells and in particular hemopoietic (Abraham, 1992). It has been shown that hemin regulates the expression of several genes, the flux of iron into the cells as well as the synthesis of hemoglobin and enzymes involved in *de novo* synthesis of heme (see Sassa for review, 1988). All these studies imply that hemin enters the cell and interacts with cellular sites at many levels. Indeed, evidence exists to indicate that tetrapyrroles, porphyrin like compounds like hemin, are transported into hemopoietic and liver cells via proteins (Mueller-Eberhard and Nikkila, 1989). These observations taken together with our most recent findings (Tsamadou et al, in preparation), that K562 cells contain several proteins that interact with hemin, prompted us to examine the kinetics of hemin's uptake in K562 cells. The data of Fig. 1 indicated that hemin enters the cells easily and accumulates in cytosol and nucleus. On light of this evidence, we examined whether hemin while in the nucleus, interacts with nuclear components. Three pieces of evidence suggest that [^{14}C]-hemin interacts with DNA: (a) First, [^{14}C]-hemin interacts with purified nuclei *in vitro*; (b) hemin forms complex with sheared cellular DNA from K562 cells and (c) DNA fragments stick only on a heme-agarose affinity column but not on a "plain" agarose column. However, these observations, are more or less preliminary in nature and add little to the specificity of the interactions between hemin and DNA. Hemin could interact with DNA either directly or via proteins. Knowing that hemin activates transcription of globin genes *in vitro* (Charney and Maniatis, 1983), we investigated whether hemin can affect possible interactions of trans-acting factors with human β-globin DNA sequences. From the data illustrated in Fig. 4 and 5, it is clear that the 131 bp region of β-globin DNA interacts like others with cytoplasmic as well as nuclear factors. Since not all of the DNA fragments obtained from the 2.0 kb 5'-end region of β-globin DNA formed complexes with trans-acting factors suggests that the trans-acting factors interact only with discrete regions of β-like globin DNA. Our observations that exogenous hemin abrogated the interactions of trans-acting factors with discrete regions of β-globin DNA indicate that either hemin interacts directly with DNA sequences or modify the trans-acting factors in a way that renders them inactive to form complexes with DNA regions. It is still quite early, to understand in molecular terms, how hemin modulates the interactions of trans-acting factors with globin DNA sequences. This phenomenon may uncover possible mechanisms via which hemin activates transcription of globin genes in K562 cells. Further analysis of this unique effect of hemin on the interaction of trans-acting factors with globin gene is currently under way in our own laboratory.

SUMMARY

Studies over the years have established that hemin (ferric protoporphyrin IX) stimulates the synthesis of embryonal, fetal but not adult type hemoglobins in human erythroleukemia K562 cells by activating transcription of the corresponding globin genes. In an effort to uncover the mechanism(s) of hemin-induced production of hemoglobins, we investigated how hemin enters inside the K562 cells and where it accumulates intracellularly. Furthermore, we investigated whether hemin interacts with intact nuclei and naked cellular K562 DNA and affects the interactions of β-globin DNA sequences with trans-acting factors. [^{14}C]-labelled hemin was used throughout this study. Our results thus far indicate that: (a) hemin enters the K-562 cells quite rapidly and accumulates mainly in cytosol. A substantial portion of hemin also accumulates in the nucleus; (b) while in the nucleus, hemin interacts with nuclear components including naked DNA; (c) hemin prevents the intaractions of trans-acting factors with discrete regions of β-globin DNA sequences including those located within the promotor. Whether or not these unique effects of hemin are related to hemin-induced hemoglobin synthesis is under current investigation.

REFERENCES

Abraham NG (1992) Heme Regulation of hematopoietic stem cell growth and development. In "*Concise Reviews in Experimental Hematology*" Ed. M.J. Murphy, Alpha Press, N. York, pp 357-373

Benz EJ, Murane MJ, Tonkonow BL, Berman BW, Mazur EM, Cavellesco C, Jenko T, Snyder EL, Forget BG, Hoffman R (1980) Embryonic-fetal erythroid characteristics of a human leukemic cell line. *Proc. Natl. Acad. Sci. USA* 77:3509

Charney P and Maniatis T (1983) Transcriptional regulation of globin gene expression in the human erythroid cell line K562. *Science* 220:1281,

Chen IT and London IM (1981) Hemin enhances the differentiation of mouse 3T3 cells to adipocyte. *Cell* 26:117

Dean A, Erard F, Schneider A, Schechter A (1981) Induction of hemoglobin accumulation in human K562 cells by hemin is reversible. *Science* 212:459

Galbraith RF, Sassa S, Kappas A (1985) Heme binding to murine erythroleukemia cells. Evidence for a heme receptor. *J. Biol. Chemistry* 260:12198

Garner M, and Revzin A (1981) A gel electrophoresis method for qualifying the binding of proteins to specific DNA regions:applications to components of the Escherichia Coli lactose operon regulatory system *Nucl. Acid Research*, 9: 3047-3060

Guarente L, Mason T (1983) Heme regulates transcription of the CYC1 gene of S. Cerevisiae via an upstream activation site. *Cell* 32: 1279-1282

Gusella JF, Weil S, Tsiftsoglou AS, Volloch V, Newman J, Keys C, Housman D (1980) Hemin does not cause commitment of MEL cells. *Blood* 56:481

Ishii OM, Maniatis G (1978) Hemin promotes rapid neurite outgrowth in cultured mouse neuroblastoma cells. *Nature*, 274: 372

Lawn RM, Efstratiadis A, O'Connell C, Maniatis T (1980) The nucleotide sequence of the human β-globin gene *Cell* 21: 647-651

Lozzio CB, Lozzio BB (1975) Human chronic myelogenous leukemia cell line with positive Philadelphia chromosome. *Blood* 45:321,

London IM, v Ernst, Fagard R, Leroux A, Levin DH, Petryshyn R (1981) Regulation of protein synthesis by phosphorylation and heme. *Cold Spring Harbor Conferences on cell proliferation*, 8: Protein phosphorylation

Maniatis T, Fritsch FF, Sambrook J (1982) *Molecular Cloning "A Laboratory Manual"* Cold Spring Harbor New York: Cold Spring Harbor Laboratory pp 180

Monette FC and Holden SA (1982) Hemin enhances the *in vitro* growth of primitive erythroid progenitor cells. *Blood* 60:527

Muller-Eberhard U and Nikkila H (1989) Transport of tetrapyrroles by proteins. *Seminars in Hematology* 26:86

Orkin SH, (1990) Globin regulation and switching: Circa 1990. *Cell* 63: 665-672

Ross J and Sautner D (1976) Induction of globin mRNA accumulation by hemin in cultured erythroleukemia cells. *Cell* 8: 513

Rowely DT, Ohlsson-Wilhelm BM, Farley BA (1985) K562 human erythro-leukemia cells demonstrate commitment. *Blood* 65: 862

Rutherford TR, Clegg JB, Weatherall DJ (1979a) K562 human leukemic cells synthesise embryonic hemoglobin in response to hemin. *Nature* 280:164, 1979

Rutherford and Weatherall DJ (1979b) Deficient heme synthesis as the cause of non-inducibility of hemoglobin synthesis in a Friend Erythroleukemia cell line. *Cell* 16: 415-423

Sassa S (1988) Heme stimulation of cellular growth and differentiation. *Seminars in Hematology* 25:312

Stryer L. (1988) Oxygen transporting proteins: Myoglobin and hemoglobin. In "*Biochemistry*" 3rd Edition, W.H. Freeman and Company, N. York, p 144,

Treisman R, Orkin SH, Maniatis T (1983) Specific transcription and DNA splicing defects in five cloned *b*-thalassemia genes *Nature* 302: 591-596

Tsamadou AI, Wong W and Tsiftsoglou AS (1992) Hemin uptake and detection of hemin binding proteins (HeBP) in human leukemia K562 cells. In "*Metal Ions in Biology and Medicine*" Vol. 2 Eds J. Anastassopoulou, Ph.Collery, JC Etienne, T. Theophanides, John Libbey Eurotext, Paris pp 145-150

Tsiftsoglou As, Bhargava KK, Rittman LS, Sartorelli AC (1981) Distribution of the inducer of differentiation Bis-Acetyl-diaminopentane in Murine erythroleukemia Cells *J. of Cell. Physiol*, 106: 419-424

Tsiftsoglou AS and Robinson SH (1985) Differentiation of leukemia cell lines: a review of focusing on murine erythroleukemia and human HL-60 cells. *Int. J. Cell Cloning* 3: 349

Tsiftsoglou AS, Wong W, Robinson SH, Hensold J (1989) Hemin increases production of β-like globin RNA transcripts in human erythroleukemia K562 cells. *Develop. Genetics* 10:311

Tsiftsoglou AS, Wong W, Tsamadou AI, Robinson SH (1991) Cooperative effects of hemin and anthracyclines in promoting terminal erythroid maturation in K562 human erythroleukemia cells. *Exp. Hematol*. 19:928

IMMUNOMODULATION BY ANTICANCER DRUGS IN THERAPEUTICS

Enrico Mihich and M. Jane Ehrke
Grace Cancer Drug Center
Roswell Park Cancer Institute
Elm and Carlton Streets
Buffalo, NY 14263 USA

INTRODUCTION

Despite the unquestionable successes achieved by cancer chemotherapy during the past 40 years, major obstacles still remain to be overcome before a large proportion of patients with the most common so-called solid tumors can be provided curative treatments. The limitations of chemotherapy are essentially related to the insufficient selectivity of antitumor activity of the drugs available and to the phenomenon of resistance. Intensive efforts are directed towards clarifying the mechanism of action of available drugs such that, based on the information accrued, greater selectivity of antitumor action may be achieved through optimization of regimens, particularly of combination chemotherapy. Likewise much work is being done to overcome and/or prevent resistance, taking advantage of recently obtained information on some of the mechanisms involved.

New areas are becoming available with potential for therapeutic exploitation due to progress in the understanding of the molecular mechanisms of regulation of cancer cells, on the one hand, and of the mechanism of host defenses operating against neoplastic diseases, on the other hand. It is reasonable to expect that eventually very specific agents will be developed affecting mechanisms of cell regulation and gene expression uniquely operating in neoplastic cells. Likewise, it is reasonable to expect that specific mechanisms of antitumor host defense can also form the basis for

NATO ASI Series, Vol. H 75
Cancer Therapy
Edited by N. D'Alessandro, E. Mihich, L. Rausa,
H. Tapiero, and T. R. Tritton

unique antitumor treatments. This latter possibility is discussed herein with emphasis on exploiting therapeutically certain modifications of the immune responses to tumors induced by certain drugs and cytokines.

Cellular products known generically as cytokines are continuously being discovered which have multiple functions ranging from the regulation of the immune system and the implementation of its cytotoxic action, to the induction of proliferation and/or differentiation of normal and tumor cells (Thomson, 1991). Thus treatments are being developed which utilize certain cytokines as cytotoxic effectors against tumors (Lotze, et al., 1991), as immunomodulating agents ultimately augmenting antitumor host responses and as stimuli to the proliferation and/or differentiation of normal cells otherwise impaired by intensive chemotherapy or by tumor-induced suppression (Metcalf and Morstyn, 1991).

The question of the relative effectiveness of treatments with modifiers of biological responses in the presence of large and of small tumor masses is not yet fully resolved. Consequently, it seems reasonable to apply agents of this type, in conjunction with cytoreduction by chemotherapy or other modalities of treatment. Moreover, combinations of immunotherapeutic and chemotherapeutic treatments may benefit from the well established concept that synergistic and more selective antitumor effects can be obtained with combinations of treatments having different mechanisms of antitumor action and different targets of limiting toxicities. It is also important to verify whether anticancer agents, when used in combination with immunotherapeutic agents, interact positively or negatively with elements of the immune response with which those agents are directly or indirectly interactive.

Immunotherapeutic treatments of cancer under clinical evaluation include the use of cytokines such as interferons, interleukin 2 (IL2) and, to a lesser extent, interleukin 1 (IL1) and tumor necrosis factor (TNF) (see DeVita, et al., 1991). Increasingly, certain of these agents are being combined with chemotherapeutic agents (e.g., IFNs plus 5FU) but often without consideration given to the potential role of immune functions. The effects, in a murine system, of Adriamycin (ADM) on certain immune functions and the interactions of this drug with IL2 and TNF are briefly discussed in this report as examples of the potential interactions between an anticancer drug with immunomodulating activity and certain cytokines with antitumor action.

RESULTS AND DISCUSSION

Immunomodulating Effects of ADM

Whereas at high doses the drug may cause immunosuppression presumably based on its cytotoxic action, at moderate doses, which are nevertheless within the therapeutic range, ADM causes a variety of immunoaugmenting effects (Table 1). These immunomodulating effects can be ascribed to an augmentation of the differentiation-activation of cells of the monocytic lineage and to an augmentation of T cell functions (Ehrke, et al., 1989b). The former effect leads to an increase of phagocytic activity, and of tumoricidal macrophages, as well as to increases in release of macrophage products such as IL1 (Mace, et al., 1988), TNF (Mace, et al., 1985) or prostaglandins (Maccubbin, et al., 1990). The T-cell functions which are modulated include: an increase in T cell mediated cytotoxicity (Tomazic, et al., 1980; Maccubbin, et al., 1992), an inhibition of a T regulatory cell (Ehrke, et al., 1984), which is different from the target of low dose cyclophosphamide (Ryoyama, et al., 1984); and an augmented production of IL2 (Ehrke, et al., 1986). It is not yet clear to what extent the increases in cytotoxic T cell responses are the consequence of the stimulation of IL2 production and/or of the inhibition of T cell down-regulatory functions; furthermore, it is also not clear to what extent, if any, the effects on T cell function are a direct drug effect or an indirect effect consequent to a primary augmentation of IL1 production by macrophage. It should be noted that effects on NK cell function vary depending on the anatomical site from which the cells tested were obtained.

TABLE 1: LIST OF THE IMMUNOMODULATORY EFFECTS OF ADM

Augments the Differentiation of Macrophages
Inhibits T Regulatory Cells
Stimulates IL-2 Production
Stimulates IL-1 Production
Stimulates TNF Production
Stimulates PGE_2 Production
Inhibits NK Cells (Spleen)
Augments NK Activity (PEC)
Stimulates LAK Cells (Tumor Bearers)
Stimulates CTL Response

The effects of ADM, summarized in Table 1, were first demonstrated in an allogeneic system and were later studied in a syngeneic system, namely the EL4 lymphoma in the C57Bl/6 mouse. In this system, implantation of as few as 10-100 EL4 cells leads to tumor-related death in 100% of the hosts. The protocol used in these studies is shown in Figure 1. ADM was given 5 days prior to tumor implantation and mice were sacrificed for immunological assays at different times after tumor implantation. Thus at sacrifice there were mice bearing large, medium or small tumors. For each response measured, the assay was carried out under previously determined optimal conditions.

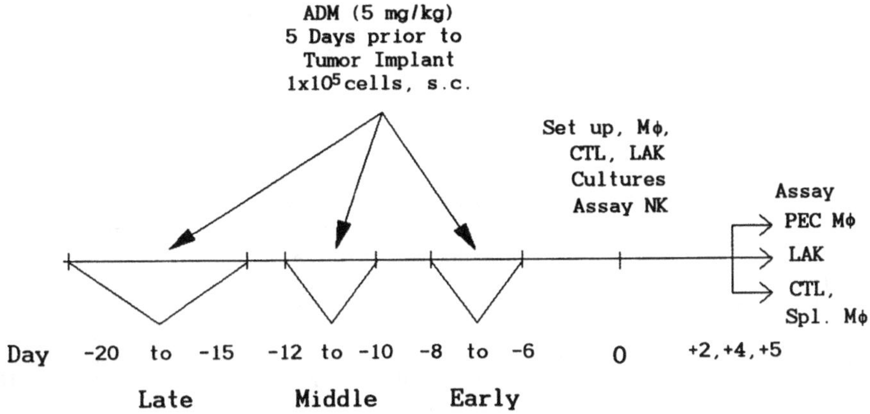

Figure 1: PROTOCOL FOR STUDIES OF THE EFFECTS OF ADRIAMYCIN (ADM) IN EL4 TUMOR BEARING MICE. C57Bl/6 mice were inoculated (s.c.) with $1x10^5$ EL4 cells at various times before the mice were sacrificed. The times were chosen so that it was possible to compare cells from animals bearing large tumors (late), small but detectable tumors (middle) and non-palpable tumors (early). At time of sacrifice spleen (S) and peritoneal (P) cells were taken for assessment. The lytic functions assessed were those of NK cells immediately on day of sacrifice or of tumoricidal macrophages (MΦ), lymphokine activated killer cells (LAK) and cytolytic T-lymphocytes (CTL) after 2, 4 or 5 days of standard stimulation culture (respectively). Lytic activity was evaluated in standard ^{51}Cr release assays using various labeled tumor cells as targets.

As shown in Table 2, most of the responses measured in untreated mice showed an initial increase, presumably related to tumor-derived antigenic stimuli, followed by severe suppression when the tumor had progressed to a large size, i.e. when the tumor had been growing in the host for a long period of time. These results suggested that the suppression of the various responses measured might be a consequence of tumor-induced suppression. This hypothesis was supported by the observation that the addition of viable tumor cells, but not x-ray killed tumor cells, to response cultures caused inhibition of the generation of allogeneic and syngeneic CTLs, LAK cells and splenic tumoricidal macrophages, without effects on NK cell activity or generation of peritoneal tumoricidal macrophages (Maccubbin, et al., 1989).

Table 2: EFFECT OF ADRIAMYCIN TREATMENT ON CYTOLYTIC RESPONSES OF CELLS FROM EL4 TUMOR BEARING MICE.

Tumor Bearing Host	ADM	CTL[b] EL4	CTL[b] P815	LAK EL4	LAK YAC	S-NK	P-NK	S-MΦ	P-MΦ
		% Specific ^{51}Cr Release							
Control	-	NA	36	51	87	20	10	36	5
(-Tumor)	+	NA	32	44	81	11	3	33	0
Early	-	14	59	49	67	17	18	25	11
	+	25	78	58	87	25	17	28	16
Middle	-	34	14	42	92	16	18	29	43
	+	29	25	60	86	27	ND	24	38
Late	-	6	1	8	47	10	7	2	0
	+	13	37	40	84	18	11	15	42

[a]Groups of C57Bl/6 mice, which had (+) or had not (-) been treated with Adriamycin (ADM, 5mg/kg) 5 days before s.c. inoculation with 5 x 10^4 cells. Separate groups of mice were sacrificed at various times after tumor inoculation (see protocol, Figure 1) and spleen and peritoneal exudate cells were taken for assessment.

[b]The lytic activities of the following effector functions were assessed: splenic (S) syngeneic and allogeneic cytotoxic T lymphocyte (CTL); splenic lymphokine activated killer (LAK); splenic and peritoneal (P) natural killer (S-NK and P-NK); splenic and peritoneal tumoricidal macrophage (S-MΦ and P-MΦ). The levels of effector (E) lytic activities are expressed as mean (4 samples) % specific ^{51}Cr release from tumor targets (T) at E:T = 50:1 as follows: syngeneic CTL and LAK against EL4; LAK and S-NK or P-NK against YAC-1, allogenic CTL, S-MΦ and P-MΦ against P815.

In most cases the response inhibition seen with cells from mice bearing large-tumor was prevented in those treated with ADM five days before tumor implantation (Table 2). That this effect was not solely due to an effect of ADM on tumor growth was indicated by the fact that immune suppression was also observed in mice bearing large EL4 lymphoma resistant to ADM (EL4/ADM) and that this suppression was absent in hosts treated with the drug. EL4/ADM is a EL4 lymphoma subline 10 fold resistant to the drug in culture, which had been developed in this laboratory and which is glycoprotein P170 positive (Ujházy, et al., 1990). Although this finding excludes that the effect of the drug in these systems is due primarily to a direct antitumor effect, it does not exclude the possibility that an indirect antitumor effect, due to the immunomodulating action of ADM, may have a role in the preservation of immune functions observed. Regardless of the precise mechanisms involved, it is likely that the immunomodulating effects observed upon treatment with ADM may be exploited in therapeutics (Mihich and Ehrke, 1991).

Therapeutic Effects of ADM in Combination With Cytokines

In order to verify the possibility that ADM-induced immunomodulation may be exploited therapeutically, experiments were carried out to see whether the drug had significant effects at an immunomodulating dose when given in therapeutic regimens, namely after EL4 implantation. Since in initial experiments no significant curative effects could be seen in this highly syngeneic system, additional investigations were carried out to see whether ADM had interactive therapeutic effects with cytokines proven to have some antitumor action by themselves. Combination with IL2 [recombinant human IL2, (DuPont)] or TNF [recombinant murine TNF, (Asahi)] were initially chosen for study because the production of both cytokines was found increased in mice treated with the drug (see above).

Numerous experiments were performed to define optimal combination regimens and then all the studies were carried out with these regimens: ADM was given at 4 mg/kg i.v. on Day 8 or on Days 1 and 8 after EL4 implantation i.p. (Day 0) and IL2 was given i.p. at 2 µg i.p. twice/day from Day 7 or 9 to Day 40. As shown in Figure 2, the regimen of ADM plus IL2 chosen based on the results of the optimization experiments had marked curative effects in the tumor model system studied.

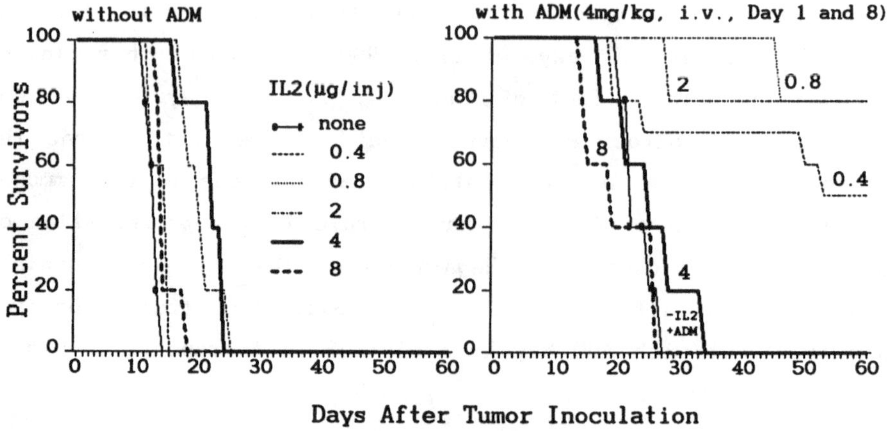

Figure 2: COOPERATIVE THERAPEUTIC EFFECTS OF ADM AND IL2. C57Bl/6 mice were inoculated (i.p.) with $5x10^4$ EL4 cells on Day 0. ADM was or was not administered as indicated. The doses of IL2 were administered i.p., b.i.d., Days 9 to 40. Survival was assessed twice daily until Day 40 and then once daily until Day 60.

Similar increases in therapeutic results were obtained when ADM, at the same dose and schedules, was used in combination with TNF 1000 units/injection i.v. given on Days 13, 16, 18, 21, and 23 after s.c. tumor inoculation (Figure 3). The question was posed as to whether the direct effects of ADM against EL4, although relatively minor per se, could have contributed to the marked therapeutic effects of the ADM combination studied. To minimize the possibility of this contribution, the regimen under study was also tested against EL4/ADM. Also in this model marked therapeutic effects were observed which were quite comparable to those seen with parent EL4 (Ehrke, et al., 1989a).

The large majority (≥90%) of the long-term survivors developed specific immunity as evidenced by their capacity to reject a re-implant of EL4 cells (Ehrke, et al., 1989a). This observation per se does not distinguish between the possibility that antitumor immune responses were instrumental in determining the curative effects of the treatments studied and the possibility that tumor immunity was a consequence of these curative effects. In order to clarify this issue further, studies in depth were carried out with these combinations to evaluate the possible active role of host defenses in determining the curative effects seen. Only the results obtained in studies of the combination of ADM plus IL2 are summarized below.

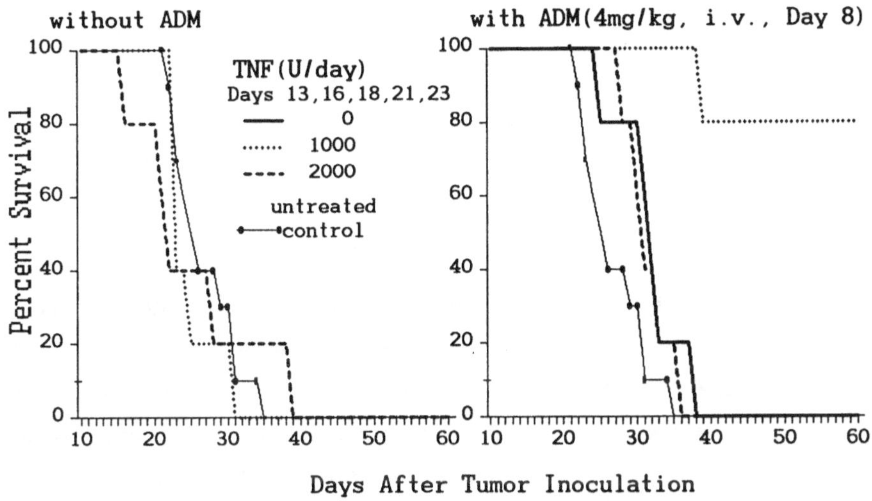

Figure 3: COOPERATIVE EFFECTS OF ADM AND TNF AGAINST EL4 LYMPHOMA. C57Bl/6 mice were inoculated (s.c.) with $5x10^4$ EL4 cells on Day 0. ADM was or was not administered as indicated. Recombinant murine tumor necrosis factor (TNF) was administered i.v. as indicated.

Role of Host Defenses in the Curative Effects of ADM Plus IL2

As shown in Table 3, the curative effects of the combination treatments were seen in non-irradiated hosts but were prevented in mice given sublethal total body irradiation (250 R/mouse) the day before EL4 inoculation. This finding suggested that those functions of host defenses which can be impaired by x-irradiation are required for the curative effects of ADM plus IL2.

Table 3: EFFECT OF SUBLETHAL, WHOLE BODY X-IRRADIATION

	Whole Body X-Irradiation			
	None		250R	
Treatment	MST[a]	LTS[b]	MST	LTS
None	13	0	12	0
IL2[c]	28	2	12	0
ADM[d]	19	0	17	0
IL2 + ADM	51	5	21	0

[a]MST = median survival time in days. 10 mice per group.
[b]LTS = long term survivors (≥ 60 days).
[c]IL2 = recombinant human interleukin 2 (DuPont) at 2 μg/injection, i.p., b.i.d., Days 9-40 (Day 0 = Day of i.p. inoculation of $5x10^4$ EL4 cells).
[d]ADM = Adriamycin, 4 mg/kg, i.v., Days 1 and 8.

In other experiments it was found that, unexpectedly, the curative effects of the combination were greater in mice implanted with $5x10^5$ EL4 cells than in those implanted with $5x10^2$ cells. Nevertheless effects comparable to those seen in mice implanted with $5x10^5$ cells were seen in mice implanted with $5x10^2$ cells when a vaccine of 10^5 non-viable EL4 cells was administered at the same time as the ADM. It, also, was found that it was possible to transfer to naive hosts anti-EL4 immunity with spleen cells from EL4 bearing donors treated with the combination regimen and that this transfer was more successful the later after initiation of treatment it was effected. Finally, and most importantly, it was shown that the curative effects of the combination regimen were totally abolished by treatment of the mice prior to tumor implantation with anti-CD8 antibody, but not by anti-NK1.1 antibody. Interestingly, pretreatment with anti-CD4 antibody partially reduced the curative effects of the combination regimen, this finding suggesting that the effects may depend, at least in part, on unimpaired T helper cell function. As a whole these results

(Table 4) strongly suggest that intact T cell functions are essential for the curative effects of the combination regimen to be seen and conversely indicate that these curative effects are indeed mediated through the positive modification of anti-EL4 immunity.

Table 4: EVIDENCE OF ACTIVE ROLE OF HOST DEFENSES IN THE THERAPEUTIC ACTION OF ADM PLUS IL2

Cures prevented by total body irradiation
Cures increased by non-viable tumor vaccine
Adoptive transfer of immune memory
Cures prevented by anti-CD8 antibody

CONCLUDING REMARKS

As discussed herein, ADM has a unique profile of multiple immunomodulating effects. In this respect, in analogy to other anticancer drugs (e.g. cyclophosphamide) which also exert unique immunomodulating effects, ADM provides an example of the until recently unsuspected specificity of the modifications of immunity potentially induced by an anticancer agent. Moreover, the results described in this report indicate that the immunomodulation caused by ADM can be exploited therapeutically in combination with a cytokine like IL2 which also has specific effects on the immune system. This example may have broader implications as current directions in therapeutics include an emphasis towards attempting to augment the therapeutic effectiveness of biological treatments and/or biological response modifications through combination with cytoreductive and/or immunomodulating chemotherapy.

ACKNOWLEDGEMENTS

The research from our laboratory reported here was supported, in part by Grants CA15142, CA24538, CA16056 and CA09072 awarded by the National Cancer Institute, Department of Health and Human Services, U.S.A. The research expertise of Drs. D.L. Maccubbin and R.L.X. Ho, our collaborators on the recent studies, are gratefully acknowledged. The authors wish to thank J. Meer and K. Schrader for their excellent assistance in data processing/illustrating/ manuscript formating and typing, respectively.

REFERENCES

DeVita VT Jr., Hellman S, Rosenberg SA (eds) (1991) Biologic Therapy of Cancer. JB Lippincott, Philadelphia PA

Ehrke MJ, Ho RLX, Mihich E (1989a) Modifications of Antitumor Immune Effectors by Adriamycin. In: Torisu M and Yoshida T (eds) New Horizons of Tumor Immunotherapy. Elsevier, Amsterdam, pp 467-475

Ehrke MJ, Maccubbin D, Ryoyama K, Cohen SA, Mihich E (1986) Correlation Between Adriamycin-induced Augmentation of Interleukin 2 Production and of Cell Mediated Cytotoxicity. Cancer Res 46: 54-60

Ehrke MJ, Mihich E, Berd D, Mastrangelo MJ (1989b) Effects of Anticancer Drugs on the Immune System in Humans. Semin Oncol 16:230-253

Ehrke MJ, Ryoyama K, Cohen SA (1984) Cellular Basis for Adriamycin-induced Augmentation of Cell Mediated Cytotoxicity in Culture. Cancer Res 44: 2497-2504

Lotze MT, Rosenberg SA (1991) Interleukin-2: Clinical Applications. In: DeVita VT Jr., Hellman S, Rosenberg SA (eds) Biologic Therapy of Cancer. JB Lippincott, Philadelphia PA, pp 59-177

Maccubbin D, Cohen S, Ehrke, MJ (1990) Indomethacin Modulation of Adriamycin Induced Effects on Multiple Cytolytic Effector Functions. Cancer Immunol Immunother 31: 373-380

Maccubbin D, Mace K, Ehrke M, Mihich E (1989) Modification of Host Antitumor Defense Mechanisms in Mice by Progressively Growing Tumor. Cancer Res 49: 4216-4224

Maccubbin D, Mace KF, Ho RLX, Ehrke MJ, Mihich E (1992) Adriamycin-modulation of Host Defenses in Tumor-bearing Mice. Cancer Res 52: 3572-3576

Mace K, Mayhew E, Mihich E, Ehrke MJ (1988) Alterations in Murine Host Defense Functions by Adriamycin or Liposome-encapsulated Adriamycin. Cancer Res 48: 130-136

Mace K, Mayhew E, Mihich E, Ehrke MJ (1985) Production of a Soluble Mediator with Tumor Lytic Activity by Adherent Peritoneal

Exudate Cells from Mice Treated with Adriamycin or Liposome Encapsulated Adriamycin. J Leukocyte Bio 38: 68

Metcalf D, Morstyn G (1991) Colony-stimulating Factors: General Biology. In: DeVita VT Jr., Hellman S, Rosenberg SA (eds) Biologic Therapy of Cancer. JB Lippincott, Philadelphia PA pp 417-444

Mihich E, Ehrke MJ (1991) Immunomodulation by Anticancer Drugs. In: DeVita VT Jr., Hellman S, Rosenberg SA (eds) Biologic Therapy of Cancer. JB Lippincott, Philadelphia PA pp 776-786

Ryoyama K, Ehrke MJ, Mihich E (1984) Induction of Suppressor T-cells in Culture II. Modification by Adriamycin. Intl J Immunopharmac 6: 521-527

Thomson A (ed) (1991) The Cytokine Handbook. Academic Press, New York

Tomazic V, Ehrke MJ, Mihich E (1980) Modulation of the Cytotoxic Response Against Allogeneic Tumor Cells in Culture by Adriamycin. Cancer Res 40: 2748-2755

Ujházy P, Chen Y, Fredericks W, Ho RLX, Baker R, Mihich E, Ehrke MJ (1990) The Relationship Between Multidrug Resistance and Tumor Necrosis Factor Resistance in an EL4 Cell Line Model. Cancer Commun 2: 181-188

DIFFERENTIAL EFFECTS OF LOW DOSES OF STRUCTURALLY DIFFERENT ANTHRACYCLINES ON IMMUNOGLOBULIN PRODUCTION BY MOUSE HYBRIDOMA B CELLS

J.-L. Teillaud and H. Tapiero[1]
Laboratoire d'Immunologie Cellulaire et Clinique, Unité INSERM 255
Institut Curie
26, rue d'Ulm
75231 Paris Cedex 05
France

Introduction.

Structurally different biological and chemical agents have been shown to block the proliferation of human and mouse tumor cells and to induce the production of molecules that are usually synthesized by terminally-differentiated cells. For instance, anti-neoplastic drugs such as Ara-C or doxorubicin (DOX) promote the irreversible induction of hemoglobin (Hb) synthesis in human erythroleukemia K562 cells with a concomitant loss of proliferative capacity (Luisi-Deluca *et al.*, 1984; Jeannesson *et al.*, 1984; Mazouzi *et al.*, 1991). Similarly, the treatment of mouse hybridoma B cells by low-doses of DOX induces the blockade of cell proliferation through an accumulation of the cells in the G2 + M phase of the cell cycle as well as a strong increase of the Immunoglobulin (Ig) production (Teillaud *et al.*, 1989), suggesting a terminal differentiation in "plasma-like" cells.

Recently, it has been reported that the growth inhibition of doxorubicin-resistant K562 cells (K562/DOX) induced by various anthracyclines is not correlated with erythroid differentiation (Mazouzi *et al.*, 1991). In addition, the increase of Ig production by hybridoma B cells induced by DOX is rapidly reversible (Teillaud *et al.*, 1989). These data raise the question of whether anthracyclines induce a "true" terminal differentiation signal or

[1] Laboratoire de Pharmacologie Cellulaire et Moléculaire, ICIG, Hôpital Paul Brousse, 94800 Villejuif, France

NATO ASI Series, Vol. H 75
Cancer Therapy
Edited by N. D'Alessandro, E. Mihich, L. Rausa, H. Tapiero, and T. R. Tritton

rather act on cell cycle and protein synthesis of hybridoma B cells through different mechanisms that do not correspond to the events occuring during the terminal differentiation of normal B cells. In fact, the mechanisms by which anthracyclines exert their cytostatic and differentiating effects are still unknown. Most of them interact with DNA and their effects on protein synthesis, transport, secretion and degradation are only poorly documented.
We have therefore examined whether a blockade of cell proliferation and an increase of Ig production by hybridoma B cells first observed with low-doses of DOX (Teillaud *et al.*, 1989) can still be achieved with structurally different anthracyclines (DOX, Pirarubicin (THP-DOX), or Aclarubicin (ACR)). In addition, we have analyzed some of the molecular events induced by these different drugs along the Ig synthesis pathway.

Blockade of cell proliferation by low doses of DOX, THP-DOX, or ACR.

In vitro exposure of mouse UN2 hybridoma B cells, that produce a monoclonal antibody (IgG2a, K) directed against sheep red blood cells (SRBC)(Ralph *et al.*, 1980) to low doses of DOX, THP-DOX, or ACR, for three days, showed an inhibition of cell proliferation without significant cell mortalility. However, different doses are required to observe this cytostatic effect depending on the anthracyclines tested. While only 5-10 ng/ml THP-DOX or ACR are sufficient to inhibit 50 % of cell proliferation, up to 50 ng/ml DOX are required to obtain a similar effect (Figure 1). These data may be related to the drug-lipophilicity and /or the charge of the anthracyclines. These characteristics may modulate the amount of drugs accumulated into the cells.
When the DNA content was analyzed by using the CellFIT/DNA cell cycle analysis software (Becton-Dickinson), a striking difference between ACR and DOX or THP-DOX-treated cells was observed. Cells treated with DOX or THP-DOX accumulated in the G2 + M phase of the cell cycle, whereas those treated with ACR exhibited a DNA content profile rather similar to that of untreated cells or showing a slight increase of the percentage of cells in the G1 phase and a decrease of the percentage of cells in the S phase (data not shown). In addition, the analysis of the Forward Angle light Scatter (FSC)

revealed an important increase in the size of DOX or THP-DOX treated cells, whereas ACR treated cells showed identical or even slightly smaller size than the untreated cells (data not shown).

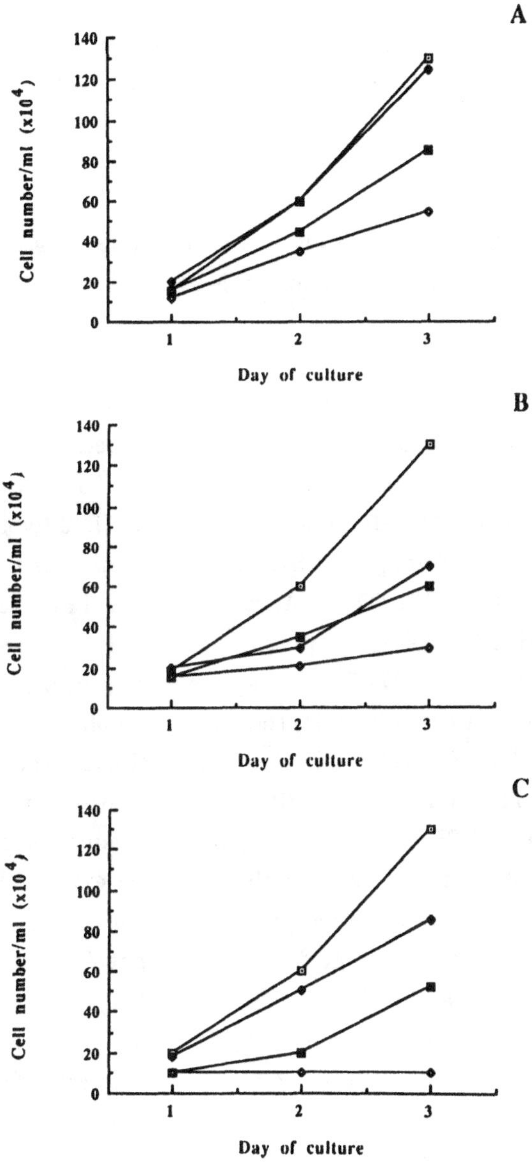

Figure 1. Effect of various doses of DOX (A)(◇ - ◇: 60 ng/ml; ■ - ■ : 30 ng/ml; ◆ - ◆: 15 ng/ml; ▫ - ▫: control), ACR (B) (◇ - ◇: 30 ng/ml; ■ - ■: 15 ng/ml; ◆ - ◆: 5 ng/ml; ▫ - ▫ control), and THP-DOX (C) (◇ - ◇: 15 ng/ml; ■ - ■: 5 ng/ml; ◆ - ◆: 2.5 ng/ml; ▫ - ▫ control) on the cell proliferation of UN2 mouse hybridoma B cells.

Thus, the inhibitory effect of low-doses of these structurally different anthracyclines on cell proliferation appears to be mediated via mechanisms that block cells at different stages of the cell cycle. DOX and THP-DOX block cells at the G2 + M phase. In contrast, ACR provokes a decrease of the percentage of cells in the S and G2 + M stages and an increase of the percentage of cells in the G1 phase. No cytotoxicity higher than 10% was detected in ACR-treated cell culture on Day 1, 2, and 3.

Cytostatic doses of DOX or THP-DOX but not of ACR induce a strong increase of Immunoglobulin production by hybridoma B cells.

Since it has been reported that these structurally-different anthracyclines induce Hb production by K562 cells, which is assumed to correspond to a terminal differentiation process (Jeannesson *et al.*, 1984), we investigated whether these drugs also affect the production of Ig by UN2 hybridoma B cells that produce IgG2a, κ directed against SRBC. Cells treated for three days with cytostatic doses of DOX, THP-DOX, or ACR were washed and tested for their IgG anti-SRBC production using a plaque assay that has been previously described (Amigorena *et al.*, 1987). This assay makes it possible to evaluate the rate of IgG secretion on a per cell basis, the size of the plaques being recorded after 30 min incubation. Figure 2 indicates that THP-DOX, as we previously reported for DOX (Teillaud *et al.*, 1989) induces a strong increase of IgG secretion. In contrast, ACR-treated cells produce plaques similar to those observed in untreated cells and even smaller in some experiments (data not shown). Thus, the blockade of cell proliferation by ACR is not associated with an increase of IgG secretion. It suggests therefore that ACR does not act via the same mechanisms as DOX or THP-DOX. These drugs may act by blocking cells in the G2 + M phase of the cell cycle or by inducing a terminal differentiation of the hybridoma B cells in "true" plasma cells. One can expect that the triggering of a terminal differentiation of hybridoma B cells by DOX and THP-DOX could be marked by a specific increase of Ig production compared to the synthesis of other proteins. In contrast, an increase of Ig production due to a blockade in the G2 + M phase of the cell cycle is likely to be accompanied by an increase of the production of many other proteins. The molecular events leading to the increased production of IgG induced by DOX or THP-treated cells were therefore analyzed.

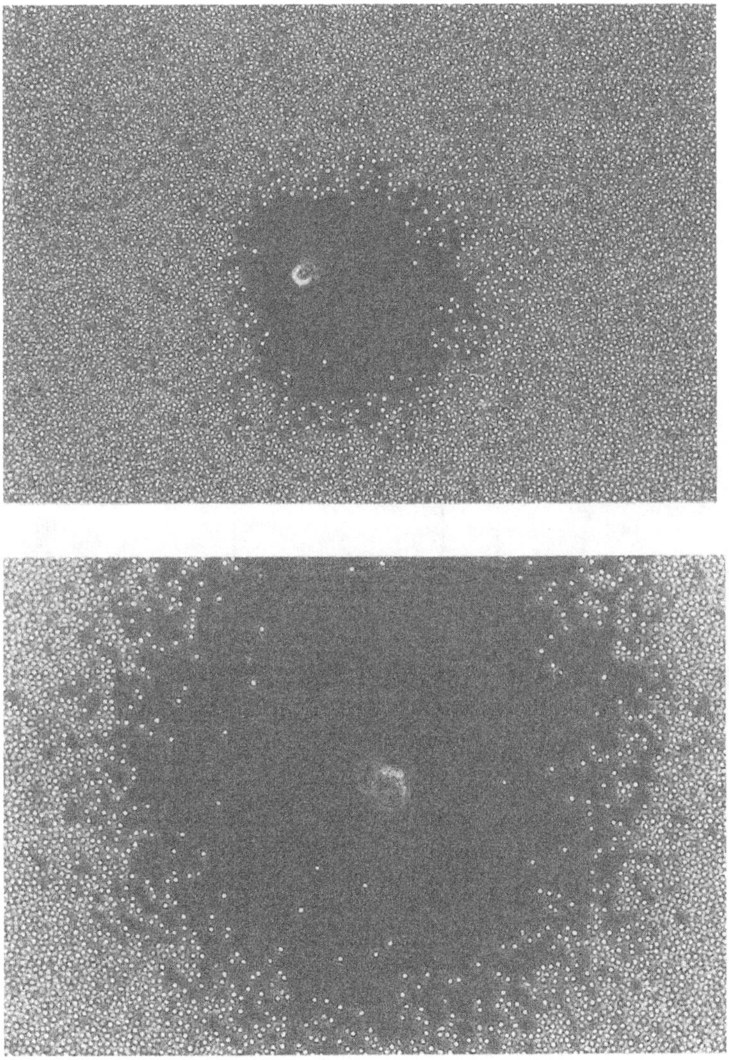

Figure 2. PFC assays of THP-DOX- or ACR-treated UN2 cells. UN2 cells were assayed in a PFC assay for their ability to lyse SRBC after 72-h incubation in the presence of ACR (5 ng/ml)(upper part) or THP-DOX (8 ng/ml)(lower part). Magnification are 400. PFC were evaluated after a 30 min incubation.

Biosynthetic labelling experiments were then performed in order to assess whether only the secretion rate of IgG was affected or if the increase of IgG secretion was due to an increase of the synthesis of the IgG heavy and light (L) chain by UN2 hybridoma B cells. Cells were incubated with cytostatic doses of either DOX, THP-DOX, or ACR for three days, washed, and labelled for 2 hours with ^{35}S-methionine. IgG present in culture supernatants and in cell extracts were then immunoprecipitated with anti-mouse IgG-coupled to Sepharose beads and analyzed on 10% polyacrylamide-SDS gels in reducing

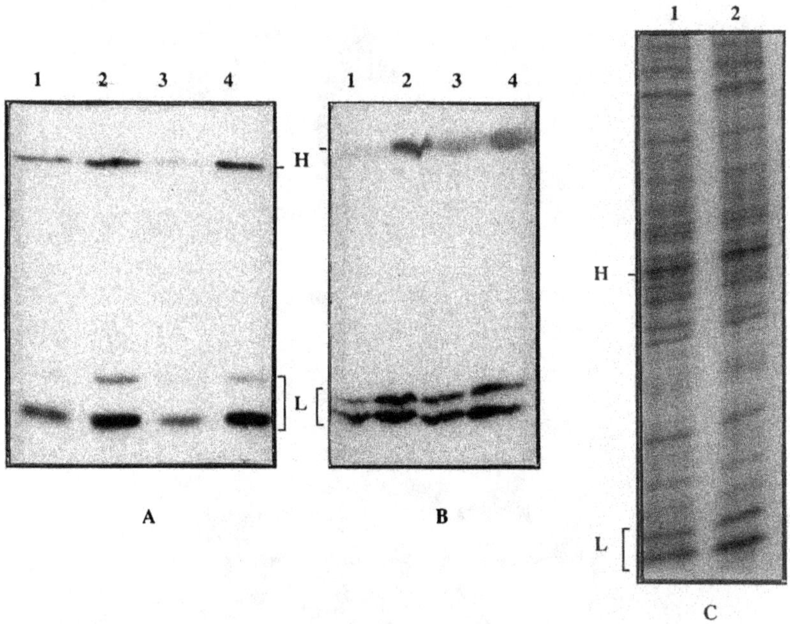

Figure 3. Analysis of protein synthesis by anthracyclines-treated UN2 cells using biosynthetic labelling with ^{35}S-methionine. UN2 cells were cultured for 3 days in presence of the different drugs or not, washed, and further incubated in fresh culture medium without drugs containing ^{35}S-methionine. (A) analysis of newly synthesized IgG present in cell culture supernatants (A) or cell extracts (B) from untreated cells (1), DOX (2), ACR (3), or THP-DOX (4)-treated cells. (B). IgG were immunoprecipitated with anti-IgG-Sepharose coupled beads and analyzed on 10% SDS-polyacrylamide gels in reducing conditions. (C) represents the comparison of radiolabelled cell homogenates of untreated (1) or THP-DOX (2)-treated cells.

conditions (Figure 3A and 3B). Both DOX and THP-DOX increased the H and L chain synthesis, whereas ACR did not affect it. The larger size of plaques is therefore related to an increase in the synthesis rate of H and L chains of IgG.

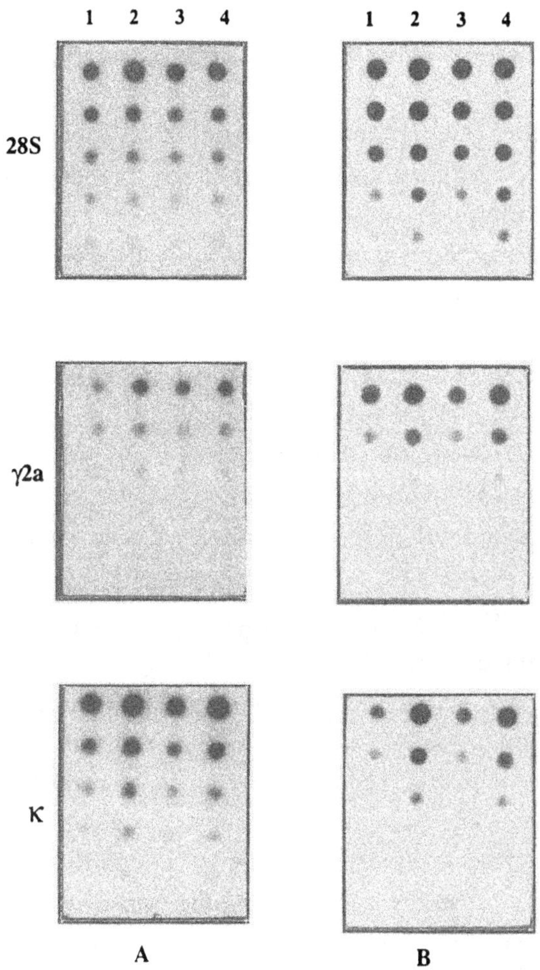

Figure 4. Analysis of 28S RNA and γ2a and κ-encoding mRNA amounts present in DOX (2), ACR (3), or THP-DOX (4)-treated or untreated (1) cells. Right pannels correspond to deposits onto nitrocellulose of total RNA extracted from the same number of cells. Left pannels correspond to the deposits of equal amounts of total RNA, as determined by A260 evaluation. Probes used in each case are indicated on the left side.

The analysis of cell extracts from THP-DOX treated cells on 10% SDS-polyacrylamide gel demonstrated a specific increase of H and L chain synthesis, as the synthesis of other protein did not appear modified compared to that of untreated cells (Figure 3C). Thus, it confirms our previous results observed with DOX-treated cells (Teillaud *et al.*, 1989). This observation argues in favor of a terminal differentiation of UN2 hybridoma B cells induced by DOX and THP-DOX rather than in favor of an increase of IgG production due to the blockade of the DOX or THP-DOX-treated cells in the G2 + M phase of the cell cycle.

The mRNA encoding for H and L chains of IgG were then analyzed (Figure 4). Total RNA were extracted from the same number of UN2 cells after 3 days of culture performed in presence of the different anthracyclines or not. Increased amounts of ribosomal 28S RNA were detected in DOX or THP-DOX treated cells, whereas ACR-treated cells exhibited 28S RNA amounts similar to those of untreated cells. This increase of 28S RNA in DOX or THP-DOX-treated cells is related to the larger size of the cells and to their accumulation in the G2 +M phase of the cell cycle. The same observation was made when γ2a and κ-encoding mRNA were analyzed (Figure 4, right pannels). Dot-blot assays of γ2a and κ-encoding mRNA were then performed after equal amounts of total RNA were deposited onto nitrocellulose filters (determined by A_{260} evaluation). Figure 4 (left pannels) shows that when the same amounts of 28S RNA are deposited (upper left pannel), γ2a and κ-encoding mRNA amounts still appear to be increased in DOX or THP-DOX treated cells relatively to the 28S RNA amounts. In contrast, ACR-treated cells exhibit γ2a and κ-encoding mRNA amounts similar to those of untreated cells. Several experiments indicated that the increase of the mRNA L chain was more pronounced than that of the mRNA H chains. It could reflect the fact that L chains are synthesized in excess compared to H chains in untreated UN2 cells, as described for many other myeloma and hybridoma cells. In conclusion, both biosynthesis labelling experiments and RNA analysis support the hypothesis of a specific increase of H and L chains expression due to a terminal differentiation of hybridoma cells induced by low-doses of DOX or THP-DOX.

Concluding remarks

The treatment of mouse hybridoma B cells for short-term periods at non-toxic doses of structurally different anthracyclines induces a differential effect depending on the drug used. DOX or THP-DOX provoke i) a strong increase of Ig secretion and synthesis per cell and ii) the accumulation of the cells in the G2+M phase of cell cycle accompanied by a significant cell enlargement. In contrast, ACR inhibit cell proliferation without modifying IgG production. This inhibition is likely to correspond to a blockade at the G1 phase of the cell cycle, as some experiments demonstrated a decrease in the percentage of the cells in the S phase. The higher amount of Ig detected after DOX or THP-DOX treatment in cell culture supernatants by ELISA (data not shown) is not due to the release of intracellular Ig by cells killed by the treatments. Cell mortality stays low in our experimental conditions and PFC assays allow the detection of viable cells still able to secrete Ig.

The enhanced synthesis of Ig by DOX or THP-DOX-treated cells may be a direct consequence of this accumulation in G2+M phase and/or of cell enlargement. Alternatively, it could be due to a terminal differentiation of hybridoma B cells. Biosynthetic labelling experiments and RNA blot analysis argue in favor of the latter hypothesis. The relative amounts of γ2a and κ-encoding mRNA are increased compared to that of 28S RNA. In addition, enhancement of protein synthesis by DOX involved only a few other proteins (Teillaud *et al.,* 1989). Dot-blot assays indicated for instance that H-2 antigen expression was also increased after DOX treatment (data not shown). In fact, the biosynthesis of most of the proteins remained unaffected by the DOX or THP-DOX treatment, as detected by the ^{35}S-methionine labelling experiments (Teillaud *et al.,* 1989)(Figure 3). However, the fact that the effect of DOX treatment on Ig production is rapidly reversible (< 8 hours)(Teillaud *et al.,* 1989) excludes a a differentiation in true "plasma-like" cells. This could account for the difference between the present report and that of Sherr *et al.* (1988). The latter indicated that only human B cell hybridomas from patients with common variable immunodeficiency (CVI) produce larger amounts of Ig when treated with the differentiation inducing agent retinoic acid.

Our data may have important consequences in term of the effects of *in vivo* anthracycline treatment : we can expect that, in contrast to ACR, low concentrations of DOX or THP-DOX arising during the clearance of the drugs will enhance the production and the secretion of Ig and possibly of cytokines produced by tumor and normal by-stander cells of the immune system of cancer patients. Ehrke *et al.* (1982) reported that mouse treatment with DOX induced a reduced *in vitro* antibody response 5 days later but an increased antibody response if the spleen cells were obtained 7 days later. This could be related to a large decrease of DOX concentrations in the animals between day 5 and day 7, leading to a situation as observed in our *in vitro* experimental system. These authors also observed a correlation between the augmentation of Interleukin 2 induced by doxorubicin and of cell-mediated cytotoxicity (Ehrke *et al.*, 1986). They reported more recently that liposome incapsulation of DOX extended the duration of DOX potentiation of murine natural killer activity (Mace *et al.*, 1988). Thus, the enhancing effect of DOX treatment due to its low serum concentrations following clearance could provoke either a clinical improvement by potentiating the activity of effector anti-tumor cells or a devastating effect by allowing tumor cells to produce more molecules such as autocrine growth factors or factors antagonizing the immune system response. The evolution of cancers could depend on the balance between these two opposite situations rather than being due strictly to the killing of tumor cells by anthracyclines such as DOX or THP-DOX.

Acknowledgments

The authors wish to gratefully acknowledge Dr. W.H. Fridman for his continuous support and helpful discussions. We thank the expert technical assistance of N. Gruel and J. Moncuit; M. Bussière and J.-P. Laborde for performing the photographic artwork. This work has been supported by INSERM and Institut Curie.

References

Amigorena S, Moncuit J, Fridman WH, Teillaud JL (1987) A sensitive method for testing the effect of immunoglobulin-binding factor on Ig secretion by hybridoma B cells. J Immunol Methods 99 : 57-64

Ehrke MJ, Cohen SA, Mihich E (1982) Selective effects of adriamycin on murine host defense systems. Immunol Rev 65 : 55-78

Ehrke MJ, Maccubbin D, Ryoyama K, Cohen SA, Mihich E (1986) Correlation between adriamycin-induced augmentation of interleukin-2 production and of cell-mediated cytotoxicity in mice. Cancer Res 46 : 54-60

Jeannesson P, Ginot L, Manfait M, Jardillier JC (1984) Induction of hemoglobin synthesis in human leukemic K562 cells by adriamycin. Anticancer Res 4 : 47-52

Luisi-Deluca C, Mitchell T, Spriggs D, Kufe DW (1984) Induction of terminal differentiation in human K562 erythroleukemia cells by arabinofuranosylcytosine. J Clin Invest 74 : 821-827

Mace K, Mayhew E, Mihich E, Ehrke MJ (1988) Alterations in murine host defense functions by Adriamycine or liposomes-encapsulated Adriamycine. Cancer Res 48 : 130-136

Mazouzi Z, Francastel C, Tapiero H, Robert-Lézénès J (1991) Growth inhibition of doxorubicin-resistant K562 leukemia cells by anthracyclines : absence of correlation with erythroid differentiation. J Cell Pharmacol 2 : 157-164

Ralph P, Nakoinz I, Diamond B, Yelton D (1980) All classes of murine IgG antibody mediate macrophage phagocytosis and lysis of erythrocytes. J Immunol 25 : 1885-1888

Sherr E, Adelman DC, Saxon A, Gilly M, Wall R, Sidell N (1988) Retinoic acid induces the differentiation of B cell hybridomas from patients with common variable immunodeficiency. J Exp Med 168 : 55-71

Teillaud J-L, Fourcade A, Huppert J, Fridman WH, Tapiero H (1989) Effects of doxorubicin on mouse hybridoma B cells : stimulation of immunoglobulin synthesis and secretion. Cancer Res 49 : 5123-5129

CHEMICAL XENOGENIZATION OF EXPERIMENTAL TUMORS BY ANTINEOPLASTIC DRUGS

P. Puccetti, U. Grohmann, R. Bianchi, L. Binaglia, M.L. Belladonna, M. Allegrucci, M.C. Fioretti
Section of Pharmacology
University of Perugia
06100 Perugia, Italy

1. Introduction

The antigenic phenotype of experimental tumors can be modified through procedures that either directly -and rather transiently, as a rule- affect the membrane structures of the cell or involve stable, often hereditary, changes in the cell biology (e.g., the genetic code). The term of *chemical xenogenization* was introduced by our group to indicate the appearance of tumor-associated transplantation antigens in murine tumors subjected to chemical treatment, and thus rendered antigenically *foreign* to the host of origin (Puccetti *et al.,* 1987). Bonmassar *et al.* (1970) had indeed found that murine leukemia cells, on repeated *in vivo* exposure to the antitumor agent dacarbazine, would become increasingly immunogenic, eventually acquiring a degree of foreignness capable of resulting in tumor cell rejection by the histocompatible host.
Therefore, chemical xenogenization indicates the induction of stable tumor variants with increased immunogenicity following exposure of the original (parental) neoplasm to different chemicals. It is now known that many drugs and chemicals, especially methylating agents, can enhance the immunogenic strength of experimental tumors, most of which will carry new transplantation antigens. In experimental models of antitumor immunotherapy, these newly acquired antigens have been successfully exploited for induction of specific, protective immunity to the otherwise poorly immunogenic antigens of the original neoplasm. The induction of cross-protective resistance to parental cells is therefore one of the most interesting features of tumor xenogenization.

NATO ASI Series, Vol. H 75
Cancer Therapy
Edited by N. D'Alessandro, E. Mihich, L. Rausa, H. Tapiero, and T. R. Tritton

2. The triazene derivatives

Many triazene derivatives, with either an imidazole or an aryl moiety, have been synthesized which possess both cytoreductive and xenogenizing properties (Fioretti *et al.*, 1981). The triazenyl-imidazole derivative DTIC (or dacarbazine) was the first compound of this class to manifest a strong ability to induce immunogenic changes in murine lymphoma cells. The experimental model most suitable to reveal xenogenization by DTIC *in vivo* is the one originally described by Bonmassar *et al.* (1970). In this model, treatment of tumor cells with DTIC leads to a progressive increase in the immunogenicity of the tumor so that previously nonimmunogenic inocula become capable of evoking a strong antitumor response which creates a state of specific resistance. In DTIC-treated animals, however, lethal tumor growth occurs since the highly immunodepressive activity of the drug prevents host antitumor responses without affecting the growth of the xenogenized line that has become resistant to the cytoreductive activity of DTIC. Therefore, the newly acquired immunogenic potential of the xenogenized tumor is only revealed by grafting the drug-treated tumor into naive, immunocompetent hosts (Bonmassar *et al.*, 1972; Riccardi *et al.*, 1978).

Studies concerning the relationship between chemical structure, cytoreductive and xenogenizing properties of triazene compounds led to the synthesis of a class of derivatives in which the imidazole ring present in DTIC was replaced by an aryl moiety (Contessa *et al.*, 1979; Nardelli *et al.*, 1984). Most of these compounds proved to be strong xenogenizing agents, thus suggesting that the imidazole ring is not mandatory for the activity. These studies also showed that dimethyl-triazenes, including DTIC, are not active *per se in vivo* but require metabolic activation, which is carried out by liver microsomal enzymes and leads to the generation of monomethyl species. When a series of dimethyl aryl-triazenes and related monomethyl compounds were assayed for induction of xenogenization *in vitro,* it was found that the dimethyl derivatives required the presence of mouse liver microsomes whereas the corresponding monomethyl compounds did not. Similar results had been obtained with DTIC and its related monomethyl derivative. Much evidence is now available that points to the strict analogy between the activity of DTIC and that of aryl-triazenes, and also suggests that the aryl-triazenes may have potential advantages over DTIC, including reduced immunotoxicity, enhanced xenogenizing properties, no need for metabolic activation, and improved physical properties. Evidence also indicates that xenogenizing properties are shared by antitumor drugs and mutagenic chemicals belonging to different classes, most notably nitrosoguanidine derivatives (Boon, 1983). Thus, although this discussion is centered on triazene derivatives that have been the subject of most studies on chemical xenogenization *in vivo,* the range of susceptible tumors and agents endowed with xenogenizing properties both *in vivo* and *in vitro* allows for the contention that chemical xenogenization is not limited to selected experimental conditions but may have broader biological significance and therapeutic implications.

3. Drug-mediated tumor antigens (DMTA)

Perhaps the most impressive feature of chemically xenogenized tumors is the acquisition of novel antigens which are responsible for the newly acquired immunogenicity. The characterization of such *drug-mediated antigens or DMTA* has been made possible through two different approaches, the first relying on reactivity of cytotoxic T lymphocytes (CTL) and the second on humoral antibody production.

The presence of DMTA on DTIC-xenogenized tumors has been investigated in a number of *in vitro* studies in which specifically sensitized CTL were tested against ^{51}Cr-labelled target DTIC lymphomas. Thus, effector CTL can be obtained in a primary *in vivo* response using, as a source of lymphocytes, spleens from animals that have rejected a DTIC tumor (Nicolin *et al.,* 1974). Similarly, specifically cytotoxic lymphocytes are generated *in vitro* in a primary response (Romani *et al.,* 1979) or in secondary responses using *in vivo* presensitized responder cells (Santoni *et al.,* 1978). The results of these studies are consistent with the hypothesis that DMTA are not detectable on the parental tumor, and indicate that DMTA are recognized by specific CD8+ cells in association with major histocompatibility complex class I determinants (Romani *et al.,* 1988).

More recently, we have resorted to the humoral antibody approach for the biochemical definition of DMTA. In particular, anti-DMTA antibodies have been used to immunoprecipitate proteins from a clone of the xenogenized L5178Y murine lymphoma line (Grohmann *et al.,* 1990). By using this antiserum, a series of tumor specific proteins mostly in the 45-80 kDa range were identified in a highly immunogenic clone (clone D). An 80 kDa component, related to retroviral gp70, was shown to possess biological activity *in vivo,* being capable of inducing a delayed-type hypersensitivity response to clone D (Grohmann *et al.,*1991) and increasing the frequency of CTL precursors in vitro to the same tumor target cells (Grohmann *et al.,* 1990; Grohmann *et al.,* 1991).

In an attempt to gain further insight into the nature of DMTA expressed by xenogenized L5178Y/DTIC cells and possibly establish the generality of clone D properties in this tumor model, we have recently investigated a series of clones derived from the same polyclonal xenogenized tumor cell line (Grohmann *et al.,* 1992). We have produced antibodies to clones B, G, I, P, Q and S that were used in immunoprecipitation studies (Table 1).

Table 1. *Immunogenicity versus expression of serological antigens in clones of the murine L5178Y/DTIC lymphoma line*

Tumor line	Immunogenicity	Antigen detected by labeling with:			
		^{35}S		^{125}I	
		p80	p45	p80	p45
Clone D	+	+	+	+	+
Clone G	+	+	+	+	+
Clone I	+	+	+	+	+
Clone P	+	+	+	+	+
Clone Q	+	+	+	+	+
Clone S	+	+	+	+	+
Clone B	+	-	+	±	±
L5178Y	-	-	-	-	+

All of these clones, which appeared to be non-tumorigenic and capable of eliciting CTL responses, displayed a pattern similar to clone D, expressing 80 and 45 kDa antigens (Table 1). Most clones also expressed variable amounts of a 30 kDa antigen, evidenced by metabolic labeling (^{35}S) and related to retroviral *gag* p30 molecules. Clone B cells, which lacked ^{35}S-labeled 80 kDa material detected by anti-clone B serum, nevertheless induced antibodies capable of immunoprecipitating this protein from the lysates of other clones and expressed 80 kDa material recognized by anti-clone G antibodies. This suggested that clone B cells contained a limited amount of 80 kDa material. In addition, an 80 kDa antigen could be detected on the surface of clone B cells by resorting to the immunoprecipitation of surface labeled material (^{125}I).

Two important points to be noted in these studies are that: a) extensive cross-reactivities among the different clones were detected by the serological approach; b) the 80 and 30 kDa antigens were both related to retrovirus-encoded structures. In several reports (Jacquemin, 1982; Apt *et al.,* 1989), the diversity of the novel antigens induced by nitrosoguanidine treatment of murine tumor cells has been associated with the extensive polymorphism of endogenous retroviral sequences, whose expression was found to be enhanced and modified following mutagen treatment. In studies of the mutagenic effect of a nitrosoguanidine derivative on cell surface-expressed molecules, Apt *et al.* (1989) reported the occurrence of antibodies that were directed against retroviral gp70 antigens in the sera of mice surviving challenge with xenogenized tumor cells. They were able to distinguish a subset of gp70 molecules with enhanced expression on xenogenized cells that could be detected by a syngeneic antiserum. The expression of another subset of gp70 molecules was induced *de novo* by mutagen treatment. In line with these data, our results indicate that the determinants recognized by anti-DMTA antibodies on clone D cells are located on molecular species related to xenotropic gp70. We hypothesize that serologically detectable p80 DMTA on clone D cells represent unique determinants rather than unique molecules. Also, our data suggest that each clonal variant of a xenogenized tumor cell line is endowed with a unique set of DMTA, although immunogenic determinants may be shared by clonal variants of the same xenogenized tumor. In addition, our more recent findings suggest that retrovirus-related sequences other than the *env* gene may be affected by triazene treatment. This was shown by the presence of serologically detectable *gag* p30-related molecules on clone D cells (and several other clones) but not on parental cells (Grohmann *et al.*, 1992). In contrast, little is presently known on the nature of the 45 kDa proteins immunoprecipitable from all xenogenized clones by means of specific antisera. Preliminary evidence suggests that these proteins may not be retrovirus-related, nor are they recognized by anti-MHC class I heavy chain monoclonal antibodies. It is likely that the precise identification of this serologically defined antigen on xenogenized cells will require purification and sequencing of the protein.

These findings are also compatible with the hypothesis that somatic mutation is the major mechanism responsible for the induction of DMTA in cells treated with triazene derivative *(vide infra)*, as has been shown to be the case for the tumor variants induced by nitrosoguanidine treatment (De Plaen *et al.*, 1988). However, our data suggest that DMTA are mostly related to products of endogenous retroviral sequences, whereas the mutant genes defined in nitrosoguaniine-treated tumors by CTL approach bear no homology to retroviral protein sequences. A possible explanation for this difference might be a higher susceptibility of retroviral sequences to the mutagenic effects of triazene derivatives. More importantly, there might be a greater ability of mutated retrovirus-related proteins to induce specific antibody responses, in addition to elicitation of a CTL response. This would imply that the serological approach preferentially detects a particular subset of mutagen-induced antigens, i.e. those related to retroviral products.

Overall, our data on DMTA characterization demonstrate that serologically detectable molecules of different size are present in all immunogenic clones of a xenogenized tumor cell line. Some of these molecules, which confer the CTL specificity, are clearly related to the products of endogenous retrovirus sequences, and display considerable antigenic cross-reactivity among different clones. This suggests that structurally abnormal, retrovirus-related proteins may act as tumor rejection antigens in xenogenized variant cells of the murine L5178Y lymphoma line.

4. Reactivity to parental antigens

In the L5178Y/DTIC tumor model system, xenogenized cells induce specific resistance to parental cells *in vivo* (Nicolin *et al.*,1976), generate T-dependent responses capable of conferring anti-parental tumor protection *in vivo* (Bianchi *et al.*,1987), and share serologically detectable *tumor-associated transplantation antigens (TATA)* with parental L5178Y cells (Romani *et al.*,1985). Although the occurrence of DMTA on the surface of xenogenized cells is believed to play a crucial role in the protection of xenogenized against parental cells, the exact mechanisms of such cross-protective immunity are poorly understood. Evidence indicates that, in addition to DMTA, the xenogenized tumor cells must share TATA with parental cells, thus suggesting that the latter antigens are presumably the target of cross-protective immune responses to the parental tumor. Though it is reasonable to hypothesize a *"helper" effect* (Keene and Forman, 1982) of DMTA leading to a stronger anti-TATA response, recent *in vitro* data failed to demonstrate an increased frequency of TATA-specific cytotoxic T lymphocyte precursors (CTLp) in response to xenogenized cells (Romani *et al.*, 1990). An example of this type of experiment is illustrated in Table 2, where limiting dilution microcultures of positively selected CD8+ lymphocytes were assayed against parental or xenogenized cells following *in vivo* priming with either tumor and restimulation *in vitro* with xenogenized cells.

Table 2. *Splenic CTLp frequencies in immunized mice**

Priming with:	Restimulation in vitro with L5178Y/DTIC	CTLp frequency to:	
		L5178Y	L5178Y/DTIC
L5178Y	-	1/54,312	1/52,128
L5178Y/DTIC	-	1/46,180	1/20,140
L5178Y	+	1/53,640	1/25,870
L5178Y/DTIC	+	1/44,320	1/5,180

* Mice received live L5178Y/DTIC or irradiated L5178Y cells 3 weeks before their use as donors of CD8+ responder cells in the limiting dilution cultures

In line with these findings, we have indeed shown that the major effector lymphocytes responsible for anti-parental tumor activity are immune Lyt-1+ CD4+ CD8⁻ cells, which are capable of passively conferring delayed-type footpad reaction *in vivo* and proliferate *in vitro* in response to parental tumor antigens (Bianchi *et al.,* 1988). Also, we have been able to demonstrate that the mechanisms of anti-parental tumor protection by xenogenized cells involve the induction of a tumor-specific delayed-type hypersensitivity (DTH) reaction initiated by CD4+ lymphocytes (Puccetti *et al.,* 1989). In a recent study, we attempted to elucidate the role of different T-cell subsets, gamma interferon (IFN-γ) production and efferent specificity in the anti-parental tumor immunity induced by xenogenized variant cells (Bianchi *et al.,* 1990). It was confirmed that tumor-specific CD4+ T lymphocytes are largely responsible for tumor-suppressive and DTH activities in our model, but IFN-γ-releasing, TATA-specific CD8+ cells could also be detected. Undoubtedly, much of our data supports the contention that parental tumor inhibition relies on specific TATA recognition by tumor-immune CD4+ and CD8+ cells, release of IFN-γ, and activation of final effective macrophages that non-specifically kill tumor target cells.

Therefore, the presence of TATA on parental cells, although not mandatory for the induction of DMTA (Fioretti *at al.,* 1980), is nevertheless necessary for the xenogenized cells to immunize effectively against parental cells. This means that the presence of DMTA makes it possible for the host to develop effective anti-TATA immunity. Although there is as yet an incomplete understanding of the mechanism(s), it seems probable that DMTA may exert an adjuvant effect through the increased function of accessory cells of the immune response and/or factors released by lymphocytes, macrophages or even the xenogenized tumor. To substantiate this hypothesis, we have recently tested a panel of murine tumors xenogenized by DTIC or a nitrosoguanidine (MNNG) derivative for production of soluble factors with lymphokine-like activity and induction of lymphokine release from naive or specifically sensitized lymphocytes (Romani *et al.,* 1989; Bianchi *et al.,* 1992). In the L5178Y tumor system, a majority of xenogenized but not parental clones were found to produce an IL-1-like factor. However, no such properties were exhibited by the xenogenized variants of P815 and L1210Ha cells, which nevertheless occasionally expressed other lymphokine activities. On examining the ability of xenogenized cells (L5178Y/MNNG, clone LM-12; P815/MNNG, clone PM-20) and their respective parental tumors to cause release of interleukin 2 (IL-2) and IFN-γ from spleen cells, we found increased lymphokine production when lymphocytes primed *in vivo* to the appropriate xenogenized tumor variants were restimulated *in vitro* with parental cells (Table 3).

Table 3. *Lymphokine production by variant-immune spleen cells*

Priming with:	Restimulation in vitro with:	Lymphokine (U/ml)	
		IL-2	IFN-γ
None	L5178Y	3	4
LM-12	L5178Y	155	62
PM-20	L5178Y	5	7
None	P815	8	4
PM-20	P815	420	82
LM-12	P815	8	5

In conclusion, these studies on anti-TATA reactivity substantiate the hypothesis that CD4+ cells and lymphokines are crucially involved in the induction of immunity to parental tumor cells by DMTA; indeed, they speak in favor of a major involvement of these factors both in the initiation of a specific response to TATA co-expressed with DMTA, and in the effector phase of the anti-parental tumor immunity mediated by tumor-specific CD4+ and CD8+ cells. In contrast, increased CTL activity to parental TATA does not seem to be a major mechanism of the protection induced by xenogenized against parental cells. These data also suggest that the mechanisms of anti-parental tumor immunity induced by MNNG-treated variants may be similar to those in the triazene xenogenization system, thus reinforcing the concept of a marked similarity between the two models (Puccetti *et al.,* 1990).

5. Mechanisms of xenogenization

The finding that immunogenic tumor variants are generated after exposure of parental cells to xenogenizing chemicals raises obvious questions as to the possible mechanisms underlying the phenomenon. It should be emphasized that several points are still controversial in this regard; however, at least in the case of triazene derivatives, the available evidence permits exclusion of some potential mechanisms. A hypothesis that has always received much attention is that emergence of immunogenic tumor sublines might result from selection of pre-existing immunogenic clones. These would not be eliminated by immunologically incompetent hosts, as the DTIC-treated animals would be expected to be. Nevertheless, much available information makes this possibility rather unlikely: thus, for instance, parental lines serially transplanted in immunodepressed mice do not give rise to immunogenic variants. Furthermore, and most importantly, DTIC xenogenization occurs in the absence of drug-induced selection, and under *in vitro* conditions. An additional possibility, in line with recent data gathered from studies with other xenogenizing chemicals, is that the triazene derivatives may activate the expression of "silent" genes and thus condition the appearance of specificities coded for by the newly activated genes. In one such epigenetic model, the interference of xenogenizing compounds with DNA resides at the level of the enzyme that methylates the base cytosine, as it is known that the extent of cytosine methylation regulates the expression of several gene functions. Immunogenic tumor variants might, therefore, have decreased levels of methylcytosine, which would increase the transcriptional activity of genes involved in the expression of immunogenicity. In this regard, we have recently shown that no detectable DNA demethylation is associated with triazene xenogenization of a murine lymphoma (Fuschiotti *at al.,* 1989). Thus, gene activation does not seem to be a major mechanism in our phenomenon. Perhaps the best explanation for drug-induced xenogenization is provided by the mutational hypothesis, which regards somatic mutation as a major factor at work in chemical xenogenization by triazenes. In this regard, it is interesting to note that retroviral DNA sequences in eukaryotic cells represent hypermutable regions, a notion which is in agreement with our finding of DMTA as *de novo* expressed, retrovirus-related peptides. Also important in this context is the recent finding that the methylation of the O^6 position of guanine in DNA plays an important role in the xenogenizing activity of triazene derivatives (Bianchi *et al.,* 1992a). Indeed, alkylation of the O^6 position of guanine makes this base prone to mispairing with a thymine, and after a subsequent cell division and DNA replication, a GC: AT transition may occur resulting in a mutational event.

6. Conclusions

Chemical xenogenization appears to be a complex phenomenon which can be induced by agents not necessarily acting through a unitary mechanism: mutagenesis of retroviral DNA domains is perhaps the major mechanism in triazene xenogenization, but other mechanisms may be involved as well. Whatever the mechanisms of xenogenization, the finding that the antigenic structure of tumor cells can be altered *in vivo* by employing appropriate treatments with antineoplastic agents or resorting to *in vitro* exposure to selected chemicals may be of relevance both for a more effective use of cytoreductive drugs and in designing new approaches to cancer immunotherapy.

On one hand, for instance, in the choice of drugs for combination chemotherapy, consideration could also be given to the xenogenizing efficiency of the drugs, or effectively xenogenizing agents might be included in the regimen. On the other hand, entirely new immunotherapeutic approaches could be developed similar to those successfully attempted in experimental models of tumor immunotherapy. Thus, for instance:

a - Tumor-bearing hosts could be treated with xenogenizing drugs in order to increase the immunogenic potential of the malignancy.

b - The host could be adoptively transferred with tumor-specific lymphocytes sensitized *in vitro* to xenogenized variants of the original neoplasm.

c - The host could be treated with with immunogenic tumor variants obtained *in vitro* by exposure of the original tumor to xenogenizing chemicals or by insertion of xenogenization genes.

d - The host could be sensitized with defined tumor antigens administered in conjunction with the immunogenic peptides resulting from xenogenization.

However, although these data are clearly a major advancement towards any practical exploitation of chemical xenogenization in the immunotherapy of human tumors, there is no doubt that much work is still needed before any firm conclusion on the possible therapeutic value of this approach can be drawn.

7. References

Apt D, Schreck J, Altevogt P: Endogenous retroviral *env* genes after N-methyl-N'-nitro-N-nitrosoguanidine treatment of mouse tumor cells: stable DNA amplifications and rearrangement. Cancer Res 49: 3966-3971, 1989.

Bianchi R, Citti L, Beghetti R, Romani L, D'Incalci M, Puccetti P, Fioretti MC: O^6-Methylguanine-DNA methyltransferase activity and induction of novel immunogenicity in murine tumor cells treated with methylating agents. Cancer Chemother Pharmacol 29: 277-282, 1992a.

Bianchi R, Fioretti MC, Grohmann U, Binaglia L, Romani L, Puccetti P: Tumor-specific L3T4+ and Lyt-2+ lymphocytes in mice primed to xenogenized tumor variants. Int J Immunopharmac 14: 915-921, 1992.

Bianchi R, Fioretti MC, Romani L, Grohmann U, Cenci E, Puccetti P: T-cell subsets, IFN-γ production and efferent specificity in anti-parental tumor immunity induced by mouse sensitization with xenogenized variant cells. Int J Cancer, 46: 653-657, 1990.

Bianchi R, Romani L, Puccetti P, Fioretti MC: Induction of tumor suppression and delayed-type footpad reaction by transfer of lymphocytes sensitized to a xenogenized tumor variant. Int J Cancer 42: 71-75, 1988.

Bianchi R, Romani L, Puccetti P, Fioretti MC: Inhibition of murine lymphoma growth by adoptive transfer of lymphocytes sensitized to a xenogenized tumor variant. Int J Cancer 40: 7-11, 1987.

Boon T: Antigenic tumor cell variants obtained with mutagens. Adv Cancer Res 39: 121-151, 1983.

Bonmassar E, Bonmassar A, Vadlamudi S, Goldin A: Antigenic changes of L1210 leukemia in mice treated with 5-(3-3'-dimethyl-1-triazeno)imidazole-4-carboxamide. Cancer Res 32: 1446-1450, 1972.

Bonmassar E, Bonmassar A, Vadlamudi S, Goldin A: Immunological alteration of leukemic cells in vivo after treatment with an antitumor drug. Proc Nat Acad Sci (Wash) 66: 1089-1095, 1970.

Contessa AR, Giampietri A, Bonmassar A, Goldin A: Increased immunogenicity of L1210 leukemia following short-term exposure to 5(3,3'-dimethyl-1-triazeno) imidazole-4-carboxamide (DTIC) in vivo or in vitro. Cancer Immunol Immunother 7: 71-76, 1979.

De Plaen E, Lurquin C, Van Pel A, Mariamé B, Szikora J-P, Wölfel T, Sibille C, Chomez P, Boon T: Immunogenic (tum-) variants of mouse tumor P815: cloning of the gene of tum-antigen P91A and identification of the tum- mutation. Proc Nat Acad Sci (Wash) 85: 2274-2278, 1988.

Fioretti MC, Nardelli B, Bianchi R, Nisi C, Sava G: Antigenic changes of a murine lymphoma by in vivo treatment with triazene derivatives. Cancer Immunol Immunother 11: 283-286, 1981.

Fioretti MC, Romani L, Bonmassar A, Taramelli D: Appearance of strong transplantation antigens in non-immunogenic lymphoma following drug treatment in vivo. J Immunopharmacol 2: 189-212, 1980.

Fuschiotti P, Fioretti MC, Romani L, Puccetti P: Lack of correlation between DNA-methylating activity and appearance of the immunogenic phenotype in clones of a murine lymphoma treated with mutagens. Cancer Immunol Immunother 29: 139-143, 1989.

Grohmann U, Puccetti P, Romani L, Binaglia L, Bianchi R, Belladonna ML, Ullrich SJ, Appella E, Fioretti MC: Immunogenic tumor variants induced by drug treatment of the L5178Y lymphoma: search for serologically defined antigens at the clonal level. Int J Cancer 52: 372-377, 1992.

Grohmann U, Romani L, Binaglia L, Fioretti MC, Puccetti P: Intrasplenic immunization for induction of humoral and cell-mediated immunity to nitrocellulose-bound antigen. J Immunol Methods 137: 9-16, 1991.

Grohmann U, Ullrich S, Mage MG, Appella E, Fioretti MC, Puccetti P, Romani L: Identification and immunogenic properties of an 80 kDa surface antigen on a drug-treated tumor variant: relationship to MuLV gp70. Eur J Immunol 20: 629-636, 1990.

Jacquemin PC: Expression of type C viral glycoproteins on P815 cells: higher expression pf Mr 70,000 glycoprotein-containing glycoprotein on immunogenic variants. Cancer Res 42: 3828-3836, 1982.

Keene JA, Forman J: Helper activity is required for in vivo generation of cytotoxic T lymphocytes. J Exp Med 155:768-782, 1982.

Nardelli B, Contessa AR, Romani L, Sava G, Nisi C, Fioretti MC: Immunogenic changes of murine lymphoma cells following *in vitro* treatment with aryl-triazene-derivatives. Cancer Immunol Immunother 16: 157-161, 1984.

Nicolin A, Bini A, Coronetti E, Goldin A: Cellular immune response to a drug treated L5178Y lymphoma subline. Nature 251: 654-655, 1974.

Nicolin A, Spreafico F, Bonmassar E, Goldin A: Antigenic changes of L5178Y lymphoma after treatment with 5-(3,3 dimethyl-1-triazeno)imidazole-4-carboxamide *in vivo.* J Natl Cancer Inst 56: 89-93, 1976.

Puccetti P, Bianchi R, Romani L, Cenci E, Fioretti MC: Delayed-type hypersensitivity to tumor antigens co-expressed with immunogenic determinants induced by xenogenization. Int J Cancer 43: 279-284, 1989.

Puccetti P, Romani L, Fioretti MC: Chemical xenogenization of experimental tumors. Cancer Metast Rev 6: 93-111, 1987.

Puccetti P, Romani L, Grohmann U, Bianchi R, Fuschiotti P, Allegrucci M, Fioretti MC: Xenogenization of experimental tumors by triazene derivatives. In: Giraldi T, Connors TA, Cartei G (eds) Tiazenes, Plenum Press, New York, pp 79-89, 1990.

Riccardi C, Fioretti MC, Giampietri A, Puccetti P, Goldin A: Growth and rejection patterns of murine lymphoma cells antigenically altered following drug treatment *in vivo.* Transplantation 25: 63-68, 1978.

Romani L, Fioretti MC, Bonmassar E: *In vitro* generation of primary cytotoxic lymphocytes against L5178Y leukemia antigenically alterated by 5-(3,3'-dimethyl-1-triazeno) imidazole-4-carboxamide *in vivo.* Transplantation 28: 218-222, 1979.

Romani L, Grohmann U, Fazioli F, Puccetti P, Mage MG, Fioretti MC: Cell-mediated immunity to chemically xenogenized tumors. I. Inhibition by specific antisera and H-2 association of the novel antigens. Cancer Immunol Immunother 26: 48-54, 1988.

Romani L, Grohmann U, Puccetti P, Rossi MA, Fioretti MC: Cell-mediated immunity to chemically xenogenized tumors. V. Failure of novel antigens to increase the frequency of tumor-specific cytotoxic T cells. Int J Immunopharmac 12: 743-749, 1990.

Romani L, Puccetti P, Fioretti MC, Mage MG: Humoral response against murine lymphoma cells xenogenized by drug treatment *in vivo.* Int J Cancer 36: 225-231, 1985.

Romani L, Puccetti P, Grohmann U, Cenci E, Mage MG, Fioretti MC: Cell-mediated immunity to chemically xenogenized tumors. IV. Production of lymphokine activity by, and in response to, highly immunogenic cells. Int J Immunopharmac 11: 537-542, 1989.

Santoni A, Kinney Y, Goldin A: Secondary cytotoxic response *in vitro* against Moloney lymphoma cells antigenically altered by drug treatment *in vivo.* J Natl Cancer Inst 60: 109-112, 1978.

REVERSAL OF DRUG RESISTANCE: SYNERGISTIC ANTI-TUMOR CYTOTOXIC ACTIVITY BY COMBINATION TREATMENT WITH DRUG AND TNF OR TOXINS

Benjamin Bonavida[1], Jeffrey T. Safrit, and Hideki Morimoto
Department of Microbiology and Immunology
UCLA School of Medicine and Jonsonn Comprehensive Cancer Center
University of California at Los Angeles, CA 90024
USA

Introduction

Current conventional therapies in the treatment of neoplastic disease include surgery, chemotherapy, and radiotherapy. However, in many instances, there is relapse with the development of metastases and highly resistant tumor cells. A major challenge remains in the development of new therapeutic modalities that can treat widespread metastases and drug resistant tumors.

Since the first anti-tumor response has been shown to play an important role in tumor rejection in experimental animals, emphasis has been placed to activate the host immune response against resistant tumors. This rationale is based on the assumption that drug resistant tumor cells are senstive to host cytotoxic mechanisms. However, this assumption has not been verified experimentally and has been challenged by one of our studies (Safrit et al., 1992) (see below). Nevertheless, several immunotherapeutic approaches are currently being tested for their therapeutic effectiveness against drug resistant tumor cells. These include LAK and TIL therapies with IL-2 (Grimm et al., 1982; Rosenberg et al., 1986), interferons and TNF alone or in combination with drugs (Krasnick et al., 1989;

[1]To whom correspondence should be addressed. The work reviewed in this manuscript was supported in part by grants from the Concern Foundation, Los Angeles, and the Boiron Foundation, Lyon, France

NATO ASI Series, Vol. H 75
Cancer Therapy
Edited by N. D'Alessandro, E. Mihich, L. Rausa, H. Tapiero, and T. R. Tritton

Bonavida et al., 1990; Mutch et al., 1989), monoclonal antibodies alone or coupled with toxins or drugs (Vitetta et al., 1991), and genetic engineering with cells that produce cytokines to boost the anti-tumor immune response.

The objective of our studies has been to develop new strategies to overcome drug resistance of tumor cells. We focused on the premise that while different cytotoxic agents exert their cytotoxic activities by different mechanisms they nevertheless may share some common pathways in the cell. If this were the case, then one might predict that combination treatment of tumor cells with two different cytotoxic agents may result in complementation and may also result in overcoming resistance to either one or both cytotoxic agents.

Do cytotoxic agents of different sources share a common pathway of anti-tumor cytotoxicity?

The mechanism of tumor cell killing by host systems as well as by chemotherapeutic drugs is not yet completely understood and has been the subject of many investigations. One major problem with chemotherapy in general is the development of resistance to the cytotoxic agents used as well as cross resistance to other unrelated drugs. Overcoming this resistance will most likely require a combination of different modalities, incuding the host's own immune response against tumors. The advent of biological response modifiers (BRM) was postulated to enhance the host immune response and to mediate anti-tumor effectors resulting in regression.

The various approaches to boost the immune response or to target immunotoxins to tumor cells are currently being experimentally and clinically investigated. A major premise of these interventions is the assumption that the tumor cells, now refractory to chemotherapy or radiation, remain sensitive to immunocytotoxic effector systems and/or toxins. Alternatively, there existed the possibility that the development of resistance to chemotherapeutic drugs would also result in resistance to cytotoxic effector cells, cytokines, microbial toxins and vice versa. We examined whether there existed a correlation between the sensitivity and resistance of tumor cells to chemotherapeutic drugs, immune effector cells or factors, and bacterial toxins.

We investigated the sensitivity of several human tumor cell lines of different histologic origin to a variety of cytotoxic systems using both short term and long term assays as described (Safrit et al.. 1992). Titrations were done in each system and the data generated were tabulated to represent relative sensitivity or resistance for each agent tested. The summary of the data analysis revealed a particular pattern shown in Table 1. The criteria for sensitivity or resistance were as follows: The (+) symbolizes sensivity to the agent with cytotoxicity of 15% or greater, while the (±) symbolizes resistance and cytotoxic values <10%. Borderline sensivity and/or resistance was symbolized by (–). Clearly, the lysis of tumor cells by cytotoxic modalities in vitro can be arranged according to a hierarchy of sensitivity and resistance.

The relative ease of obtaining lines resistance to TNF-α as compared to the difficulty in finding LAK resistant lines was striking. The pattern that emerged from these analyses of the data suggested an underlying mechanism of cytotoxicity that the various agents have in common and perhaps a shared mechanism of resistance by the tumor cell lines.

The hierarchy demonstrates, for example, that tumor cells sensitive to rTNF or NK are most likely to be sensitive to monocytes, cytotoxic drugs, toxins, and LAK cells. On the other hand, cells that are resistant to LAK or bacterial toxins will most likely be resistant to drugs, macrophages, NK, and TNF. These findings suggest that the expression of resistance to a particular agent may be linked with the simultaneous coexpressed resistance to an unrelated agent. Further, the existence of cell lines with resistance to drugs, toxins, and cross-resistance to immune effector cells, suggests that there may be a common pathway for the expression of sensitivity and resistance.

The examination of the mechanisms of cytotoxicity employed by the various agents tested reveal several effects on the biological processes of tumor cells and a few shared by different cytotoxic agents. These various cytotoxic mechanisms may not act alone but rather in concert with one another to eventually cause cell death.

These findings reveal the importance of determining the resistance patterns of tumor cells prior to therapeutic prescriptions of the cytotoxic systems and therapies. Phenomena of cross-resistance or sensitivity to therapies other than cytotoxic drugs must not be overlooked. Further, the present findings suggest that combination treatment with two agents may override the resistance to either one or

Table 1:

Susceptibility to Lysis

Tumor	rTNF	NK	Monocyte	NK + IFNa	Drugs (Adr, Cis-P)	Toxins (Dtx, Ptx)	Macrophage + IFN	LAK
U937	+	+	+	+	+	+	+	+
PA1	+	+	+	+	+	+	+	+
222	+	+	+	+	+	+	+	+
A2780	+	+	+	+	+	+	+	+
AD10(b)	+	+	+	+	(-)	+	+	+
C30(b)	+	+	+	+	(-)	+	+	+
222TR(c)	-	-	+	+	+	+	+	+
U9TR	-	-	+	+	+	+	+	+
OVC-3	-	-	-	+	+	+	+	+
322-P75	-	-	-	+	+	+	+	+
B1	-	-	-	+	+	+	+	+
C2	-	-	-	-	+	+	+	+
OVC-8	-	-	-	-	+	+	+	+
M14	-	-	-	-	+	+	+	+
RP	-	-	-	-	+	+	+	+
U251	-	-	-	+	-	-	+	+
SKOV3	-	-	-	-	-	-	+	+
RAJI	-	-	-	-	-	-	-	+
226-P59	-	-	-	-	-	-	-	-

both agents. This hypothesis was tested in two systems namely the combination of TNF-α and drugs or TNF and bacterial toxins.

Enhanced and synergistic anti-tumor cytotoxicity by combinbation of TNF-α and drugs/adriamycin in both sensitive and resistance human tumor cell lines

(A) Enhancement/cytotoxicity

We have examined several human tumor cell lines for their sensitivity and/or resistance to tumor necrosis factor (TNF) and the chemotherapeutic agent adriamycin (ADR). The tumor lines were either sensitive to both agents, resistant to one or the other, or resistant to both. Enhanced cytotoxicity was seen with the combination of TNF and ADR in each line regadless of their sensitivity or resistance patterns. Representative experiments are illustrated in Figure1 (A & B) for two resistant ovarian tumor cell lines. The synergey achieved was comfirmed by isobologram analysis (Berenbaum, 1981).

Figure 1 Cytotoxicity of the two drug resistant ovarian lines AD10 MDR$^+$ and SKOV-3 MDR$^-$ measured by the MTT assay

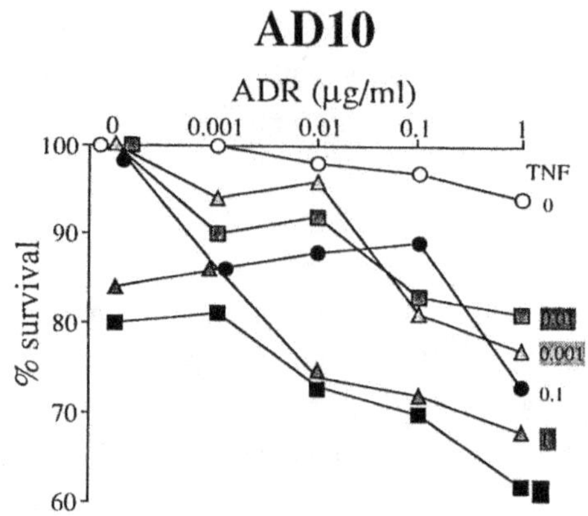

Figure 1A

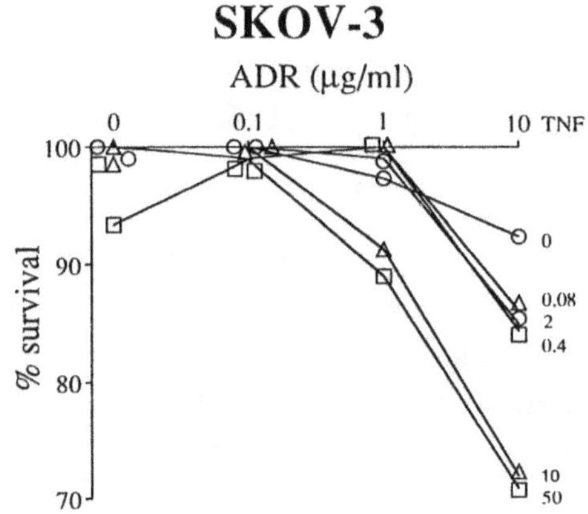

Figure 1B

(B) Mechanism of enhancement of cytotoxicity by combination treatment and overriding resistance

We examined the possible mechanisms of synergy by investigating both the role of the MDR gene and the role of TNF gene induction in resistant cells.

Because the expression of the MDR-1 gene produced by some tumor lines has been positively correlated with resistance to antineoplastic drugs, several lines were examined for the expression of the MDR phenotype by flow cytometry and northern blot analyses. One ovarian line AD10, resistant to ADR, was found to express the MDR phenotype: The AD10 line was examined for the effect of TNF and ADR used in combination on the expression of the MDR phenotype.

The effect of TNF on the influx and efflux of ADR was examined in the MDR$^-$ parental line A2780 and its MDR$^+$ line AD10. Different treatment conditions and TNF concentrations had little effect on the influx/efflux of ADR in both MDR$^-$ and MDR$^+$ lines. Further, these findings were corroborated by northern blot analyses. Incubation of AD10 with TNF, ADR or the combination had no effect on the level of MDR-1 mRNA. These results suggest that the mechanisms involved in the additive or synergistic cytotoxicity produced by combination of TNF and ADR are not due to any modification of the function or expression of the MDR phenotype by TNF (Safrit and Bonavida, 1992). Alexander et al (1987) have found that TNF synergy with anti-tumor drugs, including ADR, are targeted at DNA topiosomerase II. TNF could be acting by its suggested ability to cause DNA damage (Schmid et al., 1987) that combines with that caused by ADR to kill the cell.

We then examined resistance to TNF and whether mechanisms invoked could be involved in overcoming both TNF and drug resistance. While TNF is highly cytotoxic for some lines, it has little or no effect on others. The mechanism of TNF resistance has been associated with numerous findings such as the production of free radical scavengers like MnSOD (Wong and Goedd, 1988), over-expression of the HER2 oncogene (Hudziah et al., 1988), differential degradation of TNF in sensitive versus resistance cells (Fruehauf et al., 1991) and the induction of expression of TNF itself (Spriggs, 1987).

We examined one proposed mechanism of TNF resistance in tumor cells resulting from the induction of TNF-mRNA and TNF protein. Several tumor cell lines of different sensitivities to TNF or ADR or both were examined to delineate whether the TNF gene is regulated when combination TNF and ADR are used to overcome resistance.

Findings reported in the literature show discrepancy in the relationship between the sensitivity of tumor cells to TNF and the sensitivity to chemotherapeutic drugs. (Dolbaum et al., 1988, Neale and Matthews, 1988). We examined several cell lines of different sensitivity to TNF and/or drug and our findings demonstrate that there was no overall-correlation between the sensitivity or resistance to TNF and ADR in the various lines tested. Furthermore, the observed resistance to TNF could not be attributed to a lack of TNF receptor expression (Safrit and Bonavida, 1992).

Using a battery of tumor cell lines, we demonstrated that resistance to TNF did not necessarily correlate with either TNF secretion or the constitutive expression of the TNF mRNA in the cell lines. However, treatment of the tumor lines with rTNF for 1h resulted in the induction of varying levels of TNF mRNA expression whether the tumor lines were resistant or sensitive to TNF. Only TNF resistant lines secreted detectable levels of TNF protein either constitutively or by induction with TNF or PMA/Ionophore.

Since TNF mRNA induction correlated with resistance, the effect of TNF and ADR or reversing resistance might be due in part to downregulation of TNF mRNA by ADR. This hypothesis was tested and we found that ADR reduced the level of constitutively expressed TNF mRNA. Also, the combination of TNF and ADR reduced the level of TNF-induced mRNA in several lines (Figure 2). Downregulation of TNF mRNA occurred whether ADR was present before or after induction of the mRNA by TNF. Thus, ADR may operate by inhibiting transcription of the mRNA and also by decreasing its stability. Of interest, the TNF mRNA was not downmodulated in the MDR$^+$ AD10 line following addition of TNF and ADR even though this combination overcame resistance. The effect of ADR or TNF mRNA appears to be specific as ADR had no effect on MnSOD mRNA in one line tested.

In summary, these findings demonstrate that the resulting augmentation of and synergy cytotoxicity observed with combination treatment of TNF and ADR is of potential application in overcoming TNF and/or drug resistance. The underlying mechanisms however are complex and necessitate further studies.

Figure 2 Northern Blot Analysis of TNF-mRNA following treatment with TNF, ADR, or cycloheximide

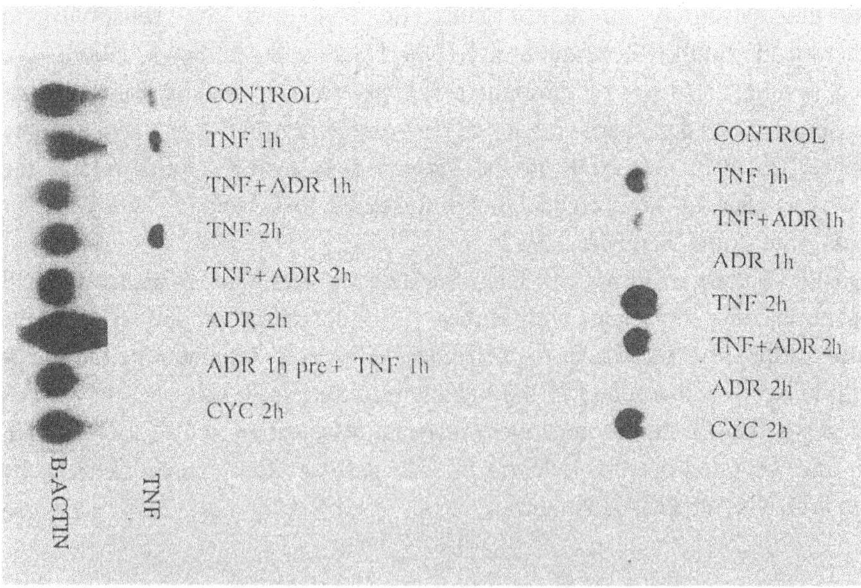

Enhancement and synergistic anti-tumor cytotoxicity by combination of TNF-α and DTX

Diphteria toxin (DTX) is an 58 Kd protein produced by C. diphteria whereby the carboxy terminal B fragment has affinity for a cell surface receptor and acts as a carrier for the amino terminal A fragment which possesses the enzymatic activity for protein synthesis inhibition (Collier and Pappenheimer, 1964). The mechanism of action of DTX constitutes a series of well defined events (Olsnes and Sandvig, 1988): the binding of DTX molecules to receptor sites on the membrane, peptide cleavage between Ag 190 and Val 191, endocytosis of toxin into vesicle compartments, conformation change of the toxin, inactivation of EF-2 by A fragment catalyzing ADP-ribosylation of a novel dipthtamide residue on EF-2, termination of protein synthesis due to depletion of active EF-2 molecules and cell death eusues.

Recent studies from our laboratory have demonstrated that DTX can also lyse target cells and induce DNA fragmentation (Chang et al., 1989) in a process reminiscent of programmed cell death or apoptosis (Wyllie el al., 1980). Of interest, TNF-α and DTX share many properties of target cell death and apoptosis. The comparison between TNF-α and DTX characteristics is summarized in Table 2.

We first investigated the effect of TNF-α and in combination with DTX in overcoming tumor cell resistance using the same approach described above for TNF-α and drugs. Three tumor cell lines of different sensitivities to TNF-α and toxin were tested and the results are shown in Figure 3. Clearly, synergistic cytotoxic activity was observed in all three lines tested and indicate that resistance to TNF-α or toxins and TNF-α is overcome by combination treatment.

Table 2 Characteristic comparison of TNF and DTX

	TNF	DTX
Molecular weight	17 kd, monomer	58 kd
Receptor binding	+ (more than one)	+
Second messenger	G-protein ?	?
Cytolysis	+	+
Kinetics of cytolysis	4-18 h	4-18 h
Protein synthesis inhibition	-	+
DNA fragmentation	+	+
Endonuclease activity	-	?
Lowering external pH	increased lysis	increased lysis
Energy poison	increased lysis	decreased lysis
Lysosomotropic agent	no effect	decreased lysis

Studies were done to investigate the mechanisms of DTX-mediated cytotoxic events. Since DTX primary activity was to inhibit protein synthesis, we examined whether protein synthesis inhibition by DTX is sufficient to inititate lysis. Figure 4 shows that while SKOV-3 protein synthesis is inhibited by DTX, the cells are not lysed. Further, the mere blocking of protein synthesis by other agents such as cyclohexamide, emetine and actinomycin D did not inhibit DTX-mediated lysis. These findings suggested that DTX mediated lysis is independent of general protein synthesis inhibition.

In a recent study, we demonstrate that target cell lysis and apoptosis mediated by DTX occurs at a step beyond the ADP ribosylation of EF-2. This was shown by using inhibitors like MESADO which blocks the biosynthesis of dipthamide and nicotinamide which competes with NAD+ on the DTX active site of EF-2, and both inhibitors inhibited DTX mediated lysis. Like the direct lysis by DTX, synergy was significantly reduced by these inhibitors. Thus, in the synergy seen with TNF-α and DTX, DTX utilizes the same pathway required for its cytolytic activity (Morimoto and Bonavida, 1992).

Figure 3: Cytotoxicity measured by the MTT assay

Sensitivities of 222, 222TR, and SKOV-3 lines to TNF and toxins

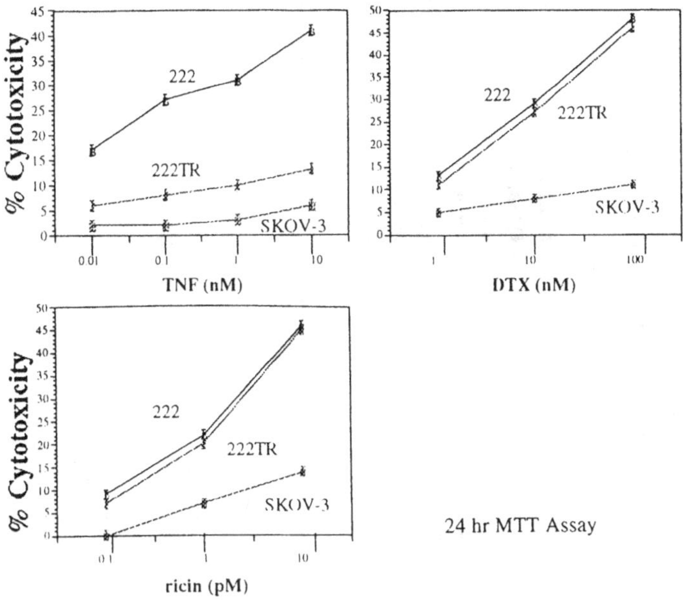

Figure 4 Cytotoxicity measured by the MTT assay

Comparison of DTX-mediated cytolytic activity and protein synthesis inhibition

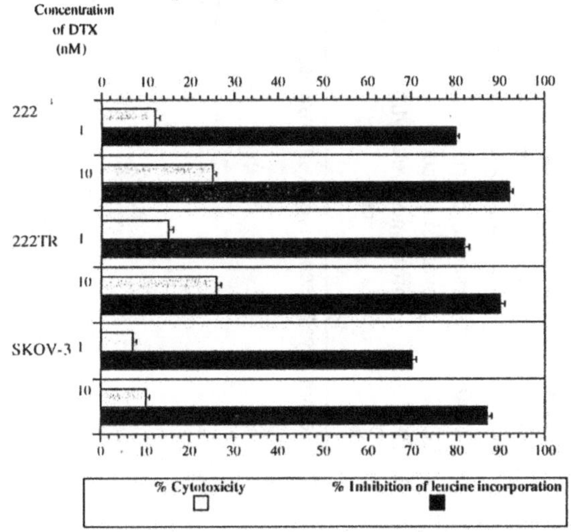

Table 3 Combination cytotoxicity of TNF and toxins

Tumor line	histology	Sensitivity*				Combination cytotoxicity	
		TNF	DTX	Ricin	Drugs (ADM, CDDP)	TNF plus DTX	TNF plus ricin
222	ovarian cancer	S	S	S	S	synergistic	additive
222TR	ovarian cancer	R	S	S	S	synergistic	additive
SKOV-3	ovarian cancer	R	R	R	R	synergistic	additive
PA-1	ovarian cancer	S	S	S	S	synergistic	additive
A2780	ovarian cancer	S	S	S	S	synergistic	additive
AD10	ovarian cancer	I	I	I	R	synergistic	additive
U937	promonomyelocytic cell	S	S	S	S	additive	additive

*Sensitivity: S, sensitive; I, intermediate; R, resistant

The plant toxin ricin was also shown to mediated target cell lysis and apoptosis. However, unlike DTX, when used in combination with TNF-α, only additive effects were achieved and not a synergistic activity. The mechanism underlying these differences between two toxins that inhibit protein synthesis and elicit DNA fragmentation is currently being investigated. A comparison showing the cytotoxicity resulting from the effect of combination treatment with TNF-α and DTX or ricin is summarized in Table 3.

Thus, our studies using TNF-α and drugs or toxins corroborate our hypothesis that cytotoxicity by different agents share common pathways. Further, the findings showing that such combinations can overcome drug resistance provide new means to explore therapeutic applications since the concentrations used in synergy are subtoxic. Also, the underlying mechanisms involved in synergy might reveal the selection of agents that can act intracellularly to initiate the cytolytic mechanism.

Alexander RB, Nelson WG, and Coffeey DS (1988) Synergistic enhancement by tumor necrosis factor of in vitro cytotoxicity from chemotherapeutic drugs targeted at DNA topoisomerases II Cancer Res 47:2403-2406

Berenbaum MC (1981) Criteria for analyzing interactions between biologically active agents. Adv Cancer Res 35:269-290

Bonavida, B., Tsuchitani, T., Zighelboim, J., and Berek, J.S (1990) Synergy is documented in vitro with low dose recombinant tumor necrosis factor, cisplatin, and doxorubicin in ovarian cancer cells. Gyn Oncol 38:333-339

Chang MP, Branhall J, Graves S, Bonavida B, and Wisnieski BJ (1989) Internucleosomal DNA cleavage precedes diptheria toxin-induced cytolysis. J Biol Chem 264:15261-15266

Collier RJ, and Pappenheimer AM (1964) Studies on the mode of action of diphteria toxin II. Effect of toxin on amino acid incorporation with free systems. J Exp Med 120:1019-1022

Dollbaum C, Creasey AA, Dairkee SH, Hiller AJ, Rudolph AR, Lin L, Vitt C, and Smith HS (1988) Specificity of tumor necrosis factor toxicity for human mammary carcinomas relative to normal mammary epithelium and correlation with response to doxorubicin Proc Nat Aca Sci USA 85:4740-4744

Fruehauf JP, Mirumaugh EG, and Sinha BK (1991) Doxorubicin induced cross-resistance to tumor necrosis factor related to differential TNF processing. J Immunotherapy 10:165-173

Grimm EA, Mazumder A, Zhang HZ, and Rosenberg SA (1982) Lymphokine activated killer cell phenomenon: Lysis of NK resistant fresh and solid tumor cells by IL-2 activated autologous human PBL. J Exp Med 155:1823-1841

Hudziak RM, Lewis GD, Shalaby RM, Eissalu TE, Aggarwal BD, Ulrich A, and Shepard HM (1988) Amplified expression of the HER/2/ERBB2 oncogene induces resistance to tumor necrosis factor in NIH 3T3 cells. Pro Nat Acad Sci 85:5102-5106

Krosnick JA, Mule JJ, McIntosh JK, and Rosenberg SA (1989) Augmentation of antitumor efficacy by the combination of recombinant tumor necrosis factor and chemotherapeutic agents in vivo. Cancer Res 49:3729-3733

Morimoto H, and Bonavida B (1992) Diptheria-toxin and Pseudomonas A toxin-mediated apoptosis: ADP-ribosylation of EF-2 is required for DNA fragmentation and cell lysis and synergy with tumor necrosis factor a. J Immunol 149:2089-2094

Mutch DG, Powell CB, Kao MS, and Collins JL (1989) In vitro analysis of the anti-cancer potential of tumor necrosis factor in combination with cisplatinum. Gyn Oncol 34:328-333

Neale ML, Matthews N (1988) Development of tumor cell resistance to tumor necrosis factor does not confer resistance to cytotoxic drugs. Eur J Cancer Clin Oncol 25: 133-137

Olsnes S, and Sandvig K (1988) How protein toxins enter and kill cells. In Immunotoxins A.E. Frankel (ed) Klumer Academic Boston p.39

Rosenberg SA, Spross P, and Lafraniere R. (1986) A new approach to the apoptive immunotherapy of cancer with tumor infiltrating lymphocytes. Science 233:1318-1320

Safrit JT, Bonavida B (1992) Sensitivity of resistant human tumor cell lines to TNF and adriamycin used in combination: Correlation between downregulation of TNF-mRNA induction and overcoming resistance. Cancer Res 52:6630-6637, 1992.

Safrit JT, Tsuchitani T, Zighelboim J, and Bonavida B (1992) Hierarchy of sensitivity and resistance of tumor cells to cytotoxic effector cells, cytokines, drugs, and toxins. Cancer Immunol. Immunotherapy 34:321-328

Schmid DS, Honing R. McGrath KM, Paul N, and Ruddle NH. (1987) Target cell DNA fragmentation is mediated by lymphotoxin and TNF Lymphokine Res 6:195-202.

Spriggs D, Imamura K, Rodriguez C, Horiguchi J, and Kufe DW (1987) Induction of tumor necrosis factor expression and resistance in a human breast tumor cell line. Proc Nat Acad Sci USA 84:6563-6566

Vitetta ES, Stone M, Amlot P, et al. (1991) Phase I immunotoxin trial in patients with B-cell lymphoma. Cancer Res 51:4052-4058

Wong GHW, Elwell JH, Oberly LW, and Goeddel DV (1989) Manganese superoxide dismutase is essential for cellular resistance to cytotoxicity of tumor necrosis factor Cell 58:923-931

Wyllie AH, Kerr JFR, and Currie AR (1980) Cell death: The significance of apoptosis Int Rev Cytol 68:251-260

IMMUNOMODULATION IN CANCER PATIENTS TREATED WITH INTERLEUKIN-2. INDUCTION OF NON-SPECIFIC AND SPECIFIC IMMUNE RESPONSES

Carlo Gambacorti-Passerini, Giorgio Parmiani, Pier Adelchi Ruffini

Division of Experimental Oncology D
Istituto Nazionale Tumori
Via Venezian, 1
20133 Milan - Italy

1. Introduction

With the discovery of the lymphokine-activated killer (LAK) cell phenomenon, a new type of tumor-killing cells that are easily obtained and expanded in the laboratory and show a broad reactivity against a variety of different tumors became available for cancer therapy (Grimm *et al.*, 1982).
After several years of experimental and clinical studies, it is now established that a proportion of advanced cancer patients, especially those with melanoma or renal metastatic cancers, respond to treatment with interleukin-2 (IL-2) plus LAK cells or even IL-2 alone (Rosenberg *et al.*, 1989). The results of those studies are summarized in Table 1. On the basis of these data and on the possible survival advantage observed in renal cancer patients treated with IL-2, such a treatment has been proposed as routine of advanced renal cancer patients (Negrier *et al.*, 1992). IL-2 has also been given the permission to be sold as effective drug in renal cancer patients in several European countries and in the USA. However, whether this treatment has to be considered more effective than other combined therapeutic modalities (e.g. IFN-α plus chemotherapy) or could be even improved by combining IL-2 with IFN-α (see Rosenberg, 1991), awaits the results of further controlled phase III clinical trials.
It is also known that adoptive immunotherapy with IL-2 at high doses is accompanied by a significant toxicity which may affect different organs and requires a considerable logistic and economic effort to be dealt with, although toxic effects reverse spontaneously upon cessation of IL-2 administration (Marolda *et al.*, 1987; Margolin *et al.*, 1989; Rosenberg *et al.*, 1989). Toxicity of IL-2 is dependent on the amount used, although subcutaneous (s.c.) or

NATO ASI Series, Vol. H 75
Cancer Therapy
Edited by N. D'Alessandro, E. Mihich, L. Rausa,
H. Tapiero, and T. R. Tritton

continuous infusion (c.i.) *vs* bolus administration has been claimed to result in a lower toxicity (West *et al.*, 1987; Atzpodien *et al.*, 1990). However, when equal amounts of IL-2 were given by bolus and by c.i., no significant difference in toxicity was found, although IL-2 given by c.i. resulted in more evident biological effects like lymphocytosis or LAK activation (Weiss *et al.*, 1989).

TABLE 1
Clinical results of treatment of advanced cancer with IL-2 plus LAK or IL-2 alone

Tumor	Treatment							
	IL-2				IL-2/LAK			
	Evaluable	CR	PR	%	Evaluable	CR	PR	%
Renal	54	4	8	22	72	8	17	35
Melanoma	42	0	10	24	48	4	6	21
Others*	33	0	0	0	21	1	3	20

CR, complete response; PR, partial response.
*Include lung, breast, colon cancers and non-Hodgkin lymphomas.

IL-2 treatment can induce *in vivo* lymphocyte proliferation, the development of broad cytotoxicity named LAK activity (Gambacorti-Passerini *et al.*, 1988) and the rise in some serum markers, such as soluble CD25 (Gambacorti-Passerini *et al.*, 1990). Unfortunately, no parameter has been consistently and strongly enough associated to clinical response in order to select responding patients from non responding ones. In conclusion, the IL-2/LAK therapy has provided numerous important findings on the mechanism of *in vivo* immune modulation although we still ignore why only a fraction of patients can respond to treatment. The low percentage of response observed in most studies (10-30%) remains the major limitation to a wider clinical use of IL-2.

We will now examine the immunobiological modifications exerted by *in vivo* IL-2 administration to see whether this treatment can activate both non-specific and tumor-specific immune reactions and whether their variable activation may explain different therapeutic effects.

2. Non-specific immunomodulation by IL-2

2.1. *Activation of natural killer (NK) cells and macrophages*

NK cells are non-T, non-B large granular lymphocytes which express a characteristic combination of differentiation antigens on their membrane, i.e. CD3-, CD16+, CD56+ (Perussia, 1991). However, functionally different subpopulations of NK cells can be distinguished on the basis of CD16 and CD56 surface expression (Perussia, 1991; Baume *et al.*, 1992). NK cells are capable of spontaneous, major histocompatibility complex (MHC)-unrestricted cytotoxic activity against a wide variety of syngeneic, allogeneic and xenogeneic malignant cells. NK cells can also mediate antibody-dependent cellular cytotoxicity (ADCC).

Mature NK cells proliferate in response to IL-2 (Perussia, 1991). Initial studies aimed at identifying the major subpopulation of peripheral blood lymphocytes (PBL) that is activated *in vivo* by IL-2 indicated that these cells have the phenotype of classical NK cells (Phillips *et al.*, 1987). It has been demonstrated that PBL exposed *in vitro* to IL-2 become able to kill otherwise NK-resistant tumor cells and such cytotoxic activity has been defined lymphokine-activated killer (LAK) cell phenomenon (Grimm *et al.*, 1982). LAK cells have been shown to be mostly IL-2-activated NK cells (Phillips and Lanier, 1986).

Demonstration that IL-2 administration can induce LAK activity *in vivo* both in patients' PBL (Gambacorti-Passerini *et al.*, 1988) and lymphocytes from cervical lymph nodes (Rivoltini *et al.*, 1990) has been provided. However, the role played by activated NK cells in destroying tumor masses during IL-2-based immunotherapy is far from clear. In fact, the immunopathology of regressing lesions in cancer patients treated with IL-2 revealed that the response is associated with T-cell and not with NK/LAK cell infiltration (Rubin *et al.*, 1989).

Macrophages, which represent the terminal differentiation stage of peripheral blood monocytes, have natural, MHC-unrestricted cytotoxic activity against tumor cell lines and can kill IgG-antibody coated target cells (ADCC) (Herberman and Ortaldo, 1981). Unstimulated human monocytes constitutively express low level of the p75 subunit (β chain of the IL-2 receptor) (Espinoza-Delgado *et al.*, 1990). The production of granulocyte-macrophage colony-stimulating factor (GM-CSF) and macrophage colony-stimulating factor (M-

CSF) has been described during IL-2 therapy (Schaafsma *et al.*, 1991) along with an increased numbers of peripheral blood granulocyte-macrophage colony forming units (CFU-GM) (Gambacorti-Passerini *et al.*, 1991). Furthermore, an increase in serum concentration of neopterin, which is a marker of macrophage activity, has been found in IL-2 treated cancer patients (Boccoli *et al.*, 1990). Immunohistochemical analysis of regressing lesions after IL-2-based immunotherapy showed a constant macrophage infiltration (Rubin *et al.*, 1989) thus suggesting a possible role for these cells in mediating tumor regression. Nevertheless, it has been suggested that IL-2-induced macrophage activation may exert a negative influence on the efficacy of cancer immunotherapies through the induction of macrophage-mediated immunosuppressive events (Lissoni *et al.*, 1991).

2.2. *Cytokine cascade*

The work of several groups has indicated that soluble mediators, distinct from IL-2, are released during immunotherapy with IL-2 either from exogenously provided or endogenously activated LAK cells or from other host's cells as well (Gemlo *et al.*, 1988; Boccoli *et al.*, 1990). This "cytokine cascade" is thought to play important roles in both the therapeutic and toxic effects (see Parmiani *et al.*, 1991). In fact, lymphocytes exposed to IL-2 *in vitro* can express genes coding for TNF-α, TNF-β, IFN-γ, IL-4, IL-5, IL-6, GM-CSF (Belldegrun *et al.*, 1989; Mazzocchi *et al.*, 1990). On one hand, such cytokines can contribute to the capillary leak syndrome which determines most toxic symptoms, on the other hand the tumor cytotoxic activity of TNF and/or IFN-γ may facilitate the distruction of neoplastic cells (Mazzocchi *et al.*, 1990).

2.3. *Activation of eosinophils*

Eosinophilia is often described in patients receiving IL-2 and its peak depends on the schedule of IL-2 administration. Silberstein *et al.* (1989) first described the appearance of hypodense (i.e. activated) eosinophils in peripheral blood of patients treated with IL-2. Eosinophils obtained from the peripheral blood of patients receiving IL-2 s.c. showed an enhanced direct or ADCC activity against tumor cells when compared with eosinophils from healthy donors or from pretreated patients (Table 2). It was also found that IL-5 contained in the patients' sera and released during IL-2 therapy was responsible of eosinophils activation and antitumor activity. Although IL-5 is known to be released by T

cells, IL-5 mRNA was also detected in eosinophils, suggesting a possible autocrine loop (Rivoltini *et al.*, unpublished data).
Thus our results support a role of eosinophils in the anti-tumor effects induced *in vivo* by IL-2. These granulocytes could exert either a cytotoxic function against tumor cells or, *via* the expression of MHC class II molecules, play an accessory function as antigen presenting cells for T lymphocytes.

TABLE 2
Direct anti-tumor activity of eosinophils from three cancer patients given IL-2 s.c.

Tumor	Eosinophils/	Patient N.		
target	target cells ratio	1	2	3
N592	50/1	51*	49	31
	25/1	40	33	31
	12/1	32	24	25
K562	50/1	27	29	28
	25/1	12	17	9
	12/1	1	9	3

N592 is a lung carcinoma and K562 an erithroleukemia line, respectively.
*Percentage of ^{51}Cr-release in a 18 hr assay.

3. Activation of tumor-specific T cells

Animal models have shown that tumor-specific T lymphocytes are by far more effective then LAK cells in the treatment of immunogenic tumors (see Rosenberg *et al.*, 1989; Rodolfo *et al.*, 1990).
Therefore the characterization of this type of cells in human tumors, and the effect of IL-2 treatment on them is of paramount importance for both the immune monitoring of IL-2 treated patients and for the design of innovative treatment modalities.

3.1. *Immune monitoring of IL-2 treated patients*

So far few studies have analyzed the *in vivo* effects of IL-2 on the ability of patient lymphocytes to kill autologous tumor cells. These studies have usually shown that the use of moderate or even low doses of IL-2 can induce an anti-autologous tumor cytotoxic activity in the patients' peripheral blood (Gambacorti-Passerini *et al.*, 1988) or cervical node lymphocytes (Rivoltini *et al.*, 1990). This activity, however, was constantly accompanied by the usual MHC-unrestricted cytotoxic activity (LAK activity) commonly induced by IL-2. Therefore, it is difficult in this setting to establish the existence of tumor-specific lymphocytes and the effect of IL-2 treatment of them. The use of high doses of IL-2 could paradoxically be suppressive for the generation of antigen-specific lymphocytes (Wiebke *et al.*, 1988).

The relative role of tumor-specific T cells and of activated NK cells ("LAK cells") in causing the tumor responses observed in IL-2 treated patients is highly debated (Parmiani, 1990). In experimental systems, NK cells, although responsible for most of the toxicity, contributed minimally to the therapeutic response to adoptive immunotherapy, while tumor-specific T lymphocytes were responsible for the cure of the IL-2 treated animals (Peace *et al.*, 1989).

3.2. *Tumor Infiltrating Lymphocytes (TIL)*

Early animal and clinical studies suggested that the use of lymphocytes obtained directly from the tumor mass (TIL) could increase the therapeutic activity of IL-2 (see Rosenberg, 1991).

Even if other studies only partially confirmed this finding (Dillman *et al.*, 1991; Parmiani *et al.*, 1992), preclinical studies analyzing the specificity of TIL indicated that these effectors can contain a frequency of tumor-specific lymphocytes higher than LAK cells (Muul-Mesler *et al.*, 1987; Itoh *et al.*, 1988).

These studies therefore were useful in defining the existence of tumor-specific lymphocytes in human tumors, thus confirming earlier reports (Fossati *et al.*, 1984). They however fell short of identifying the molecules recognized by these functionally-defined "tumor-specific lymphocytes".

3.3. *Molecular analysis of tumor-specific antigens*

This latter part belongs rather to the future approaches to tumor immunotherapy than to past achievements. The knowledge of the molecules recognized by tumor-specific lymphocytes would change dramatically the research approach in tumor immunology.

The gap between functionally defined and molecularly characterized antigens is now being filled. We now know a set of new genes, named MAGE, which code for melanoma antigens recognized by autologous lymphocyte clones (Van der Bruggen *et al.*, 1991). This knowledge will permit to specifically target molecules expressed only by the tumor cells, with the hope for fewer side effects and increased therapeutic ratio. In this context, IL-2 will still be a component of the treatment, but it will be used to boost a specific and known response rather than a non-specific one. IL-2 dosages will probably be lower than those presently employed.

Although specifically expressed on tumor cells, little is known on the biological activity of the MAGE gene family and its relationship to the process of neoplastic transformation. Particularly, it is not known whether they are merely associated with malignant transformation or play an active role in it; in other words, how easily the tumor cells can loose the expression of MAGE proteins and retain their malignant phenotype, thus escaping a possible anti-MAGE immune response. In parallel, the products of different oncogenes such as the mutated ras oncogenes (Jung and Schluesener, 1991) or some fusion proteins present in chronic myeloid leukemia and in acute promyelocytic leukemia (Chen *et al.*, 1992; Gambacorti-Passerini *et al.*, 1993), have been found to be specifically recognized by human or murine lymphocytes. These results may open the way to the future targeting of gene products directly involved in the malignant transformation of normal cells.

4. References

Atzpodien J, Korfer A, Franks CR, Poliwoda H, Kirchner H (1990) Home therapy with recombinant interleukin-2 and interferon-α2b in advanced human malignancies. Lancet 335: 1509-1512

Baume DM, Robertson MJ, Levine H, Manley TJ, Schow PW, Ritz J (1992) Differential responses to interleukin-2 define functionally distinct subsets of human natural killer cells. Eur J Immunol 22: 1-6

Belldegrun A, Kasid A, Uppenkamp M, Topalian SL, Rosenberg SA (1989) Human tumor infiltrating lymphocytes. Analysis of lymphokine mRNA expression and relevance to cancer immunotherapy. J Immunol 142: 4520-4526

Boccoli G, Masciulli R, Ruggeri EM, Carlini P, Giannella E, Montesoro G, Mastroberardino G, Isacchi G, Testa U, Calabresi F, Peschle C (1990) Adoptive immunotherapy of human cancer: the cytokine cascade and monocyte activation following high-dose interleukin-2 bolus treatment. Cancer Res 50: 5795-5800

Chen W, Peace DJ, Rovira DK, You SG, Cheever MA (1992) T-cell immunity to the joining region of $p210^{BCR-ABL}$ protein. Proc Natl Acad Sci (USA) 89: 1468-1472

Dillman RO, Oldham RK, Barth NB, Cohen RJ, Mino DR, Birch R, Yannelli JR, Maleckar JR, Sferruzza A, Arnold J, West WH (1991) Continuous interleukin-2 and tumor-infiltrating lymphocytes as treatment of advanced melanoma. Cancer 68: 1-8

Espinoza-Delgado I, Ortaldo JR, Winkler-Pickett R, Sugamura K, Varesio L, Longo DL (1990) Expression and role of p75 IL-2 receptor on human monocytes. J Exp Med 171: 1821-1832

Fossati G, Taramelli D, Balsari A, Bogdanovich G, Andreola S, Parmiani G (1984) Primary but not metastatic human melanomas expressing DR antigens stimulate autologous lymphocytes. Int J Cancer 33: 591-597

Gambacorti-Passerini C, Grignani F, Arienti F, Pandolfi PP, Pelicci PG, Parmiani G (1993) Human CD4 lymphocytes specifically recognize a peptide representing the fusion region of the hybrid protein pml/RARα present in acute promyelocitic leukemia cells. Blood: in press

Gambacorti-Passerini C, Hank J, Borchert A, Moore K, Malkovska V, Sondel P (1991) In vivo effects of multiple cycles of recombinant interleukin-2 on peripheral granulocyte-macrophage hematopoietic progenitors circulating in the blood of cancer patients. Tumori 77: 420-422

Gambacorti-Passerini C, Radrizzani M, Marolda R, Belli F, Sciorelli G, Galazka A, Schindler JD, Cascinelli N, Parmiani G (1988) In vivo activation of lymphocytes in melanoma patients receiving escalating doses of recombinant interleukin-2. Int J Cancer 41: 700-706

Gambacorti-Passerini C, Rovelli F, Ghezzi R, Traversari C, Belli F, Parmiani G (1990) Serum CD25 levels in patients treated with rIL2 and LAK. Correlation with in vivo immune activation. J Immunol Res 2: 19-24

Gemlo BT, Palladino MA, Jaff HS, Espevik TP, Rayner AA (1988) Circulating cytokines in patients with metastatic cancer treated with recombinant interleukin-2 and lymphokine-activated killer cells. Cancer Res 48: 5864-5867

Grimm EA, Mazumder A, Zhang HZ, Rosenberg SA (1982) Lymphokine-activated killer cell phenomenon: lysis of natural killer-resistant fresh solid tumor cells by interleukin-2-activated autologous human peripheral blood lymphocytes. J Exp Med 155: 1823-1841

Herberman RB, Ortaldo JR (1981) Natural killer cells: their role in defenses against disease. Science 214: 24-30

Itoh K, Platsoucas CD, Balch CM (1988) Autologous tumor-specific cytotoxic T lymphocytes in the infiltrate of human metastatic melanomas. J Exp Med 168: 1419-1441

Jung S, Schluesener HJ (1991) Human T lymphocytes recognize a peptide of single point-mutated, oncogenic ras proteins. J Exp Med 173: 273-276

Lissoni P, Tisi E, Brivio F, Barni S, Rovelli F, Perego M, Tancini G (1991) Increase in soluble interleukin-2 receptor and neopterin serum levels during immunotherapy of cancer with interleukin-2. Eur J Cancer 27: 1014-1016

Margolin K, Rayner AA, Hawkins MJ, Atkins MB, Dutcher JP, Fisher RI, Weiss GR, Doroshow JH, Jaffe HS, Roper M, Parkinson DR, Wiernik PH, Creekmore SP, Boldt DH (1989) Interleukin-2 and lymphokine-activated killer (LAK) cell therapy of solid tumors. Analysis of toxicity and management guidelines. J Clin Oncol 7: 486-498

Marolda R, Belli F, Prada A, Villani F, Gambacorti-Passerini C, Galazka A, Parmiani G, Cascinelli N (1987) A phase I study of recombinant interleukin-2 in melanoma patients. Toxicity and clinical effects. Tumori 73: 575-584

Mazzocchi A, Anichini A, Castelli C, Sensi M, Poli F, Russo C, Parmiani G (1990) T lymphocytes can mediate lysis of autologous melanoma cells by multiple mechanisms: evidence with a single T cell clone. Cancer Immunol Immunother 32: 13-21

Muul-Mesler L, Spiess PJ, Director EP, Rosenberg SA (1987) Identification of specific cytolytic immune responses against autologous tumor in humans bearing malignant melanoma. J Immunol 138: 989-995

Negrier S, Mercatello A, Coronel B, Ravaud A, Merrouche Y, Bret M, Blay JY, Lasset C, Favrot M, Moskovtchencko JF, Philip T (1992) Interleukin-2 in renal cell carcinoma. Adv Cancer Ther 2: 4-8

Parmiani G (1990) An explanation of the variable clinical response to interleukin-2 and LAK cells. Immunol Today 11: 113-116

Parmiani G, Anichini A, Rivoltini L, Arienti F, Belli F, Cascinelli N (1991) Immunotherapy of cancer with IL-2 and lymphocytes: possible reasons for the variable clinical response. In: Frontiers in Autoimmune Diseases and Cancer Immunotherapy, Dammacco F ed, Edi-ermes, Milano, 307-323

Parmiani G, Arienti F, Rivoltini L, Belli F, Cascinelli N (1992) Human tumor infiltrating lymphocytes in the treatment of cancer. Forum 2: 412-419

Peace DJ, Cheever MA (1989) Toxicity and therapeutic efficacy of high-dose interleukin-2. In vivo infusion of antibody to NK-1.1 attenuates toxicity without compromising efficacy against murine leukemia. J Exp Med 169: 161-173

Perussia B (1991) Lymphokine-activated killer cells, natural killer cells and cytokines. Curr Opin Immunol 3: 49-55

Phillips JH, Gemlo BT, Meyers WW, Rayner AA, Lanier LL (1987) In vivo and in vitro activation of natural killer in advanced cancer patients undergoing combined recombinant interleukin-2 and LAK cell therapy. J Clin Oncol 5: 1933-1941

Phillips JH, Lanier LL (1986) Dissection of the lymphokine-activated killer phenomenon. Relative contribution of peripheral blood natural killer cells and T lymphocytes to cytolysis. J Exp Med 164: 814-825

Rivoltini L, Gambacorti-Passerini C, Squadrelli-Saraceno M, Grosso MI, Cantu` G, Molinari R, Orazi A, Parmiani G (1990) In vivo IL-2 induced activation of lymphokine-activated killer cells and tumor cytotoxic T cells in cervical lymph nodes of patients with head and neck tumors. Cancer Res 50: 5551-5557

Rodolfo M, Salvi C, Bassi C, Parmiani G (1990) Adoptive Immunotherapy of a mouse colon carcinoma with recombinant interleukin-2 alone or combined with lymphokine-activated killer cells or tumor-immune lymphocytes. Cancer Immunol Immunother 31: 28-36

Rosenberg SA (1991) Immunotherapy and gene therapy of cancer. Cancer Res (Suppl) 51: 5074-5079

Rosenberg SA, Lotze MT, Yang JC, Aebersold PM, Marston Linehan W, Seipp CA, White DE (1989) Experience with the use of high-dose interleukin-2 in the treatment of 652 cancer patients. Ann Surg 210: 474-485

Rubin JT, Elwood LJ, Rosenberg SA, Lotze MT (1989) Immunohistochemical correlates response to interleukin-2 based immunotherapy in humans. Cancer Res 49: 7086-7092

Schaafsma MR, Falkenburg JH, Landegent JE, Duinkerken N, Osanto S, Ralph P, Kaushansky K, Wagemaker G, Van Damme J, Willemze R, Fibbe WE (1991) In vivo production of interleukin-5, granulocyte-macrophage colony-stimulating factor, macrophage colony-stimulating factor and interleukin-6 during intravenous administration of high-dose interleukin-2 in cancer patients. Blood 78: 1981-1987

Silberstein DS, Schoof DD, Rodrick ML, Tai PC, Spry CJ, David JR, Eberlein TJ (1989) Activation of eosinophils in cancer patients treated with IL-2 and IL-2-generated lymphokine-activated killer cells. J Immunol 142: 2162-2167

Van der Bruggen P, Traversari C, Chomez P, Lurquin C, De Plaen E, Van den Eynde B, Knuth A, Boon T (1991) A gene encoding an antigen recognized by cytolytic T lymphocytes on a human melanoma. Science 254: 1643-1647

Weiss GR, Margolin K, Aronson FR, Sznol M, Atkins MB, Guclap R, Fisher RI (1989) A randomized phase II trial of continuous infusion interleukin-2 or bolus injection rIL-2 plus lymphokine-activated killer cells for advanced renal cell cancer. Proc Am Soc Clin Oncol 8: 131

West WH, Tauer KW, Yannelli JR, Marshall GD, Orr DW, Thurman GB, Oldham RK (1987) Constant-infusion recombinant interleukin-2 in adoptive immunotherapy of advanced cancer. N Engl J Med 316: 898-905

Wiebke EA, Rosenberg SA, Lotze MT (1988) Acute immunologic effects of interleukin-2 in cancer patients: decreased delayed type hypersensitivity response and decreased proliferative response to soluble antigens. J Clin Oncol 9: 1440-1449

POTENTIAL ROLE OF TUMOR CELL ANTIGEN MODULATION IN CANCER IMMUNOTHERAPY

F. Guadagni
Laboratory of Clinical Pathology
Regina Elena Cancer Institute
Viale Regina Elena, 291
00161 Rome
Italy

Since the development of hybridoma technology (Kohler and Milstein 1975), investigators have focused their attention on the development of different approaches for the utilization of this technology in the detection and treatment of human cancer. The use of MAbs for clinical management of neoplastic patients includes clinical diagnosis using serum assays, immunocytopathological analyses of effusions or fine-needle aspiration specimens, immunoscintigraphy and, with additional development, site-directed immunotherapy. Their diagnostic application may provide a good opportunity to preselect patients for the immunotherapy on the basis of the localization of the MAb to tumor lesions through imaging studies. One advantage initially suggested for MAb-based immunotherapy was the selective nature by which an antibody conjugated with a drug, toxin, or radionuclide would target a tumor-positive carcinoma cell population. Several clinical trials utilizing MAbs conjugated with radionuclides have shown preferential localization to human carcinomas (Colcher *et al*, 1987a, b; Colcher *et al*, 1990; Schlom *et al*, 1989, 1991). Among the various MAbs generated, MAb B72.3 has been extensively used in clinical trials either for antigen identification in sera, or for tumor localization in carcinoma patients. Monoclonal antibody B72.3 was developed immunizing mice with a membrane-enriched fraction of a human carcinoma metastasis. It is a murine IgG_1, and the reactive antigen is a high molecular

NATO ASI Series, Vol. H 75
Cancer Therapy
Edited by N. D'Alessandro, E. Mihich, L. Rausa,
H. Tapiero, and T. R. Tritton

weight mucin-like glycoprotein, termed TAG-72 (tumor-associated glycoprotein-72). B72.3 has been reacted with a spectrum of adult and fetal human tissues using an immuno-histochemical technique to evaluate the expression of the reactive TAG-72 antigen. TAG-72 was found in a wide variety of human adenocarcinomas, including breast, colon, ovary, pancreas, stomach, endometrial and non-small cell lung carcinoma (Thor *et al*, 1986), suggesting that the TAG-72 antigen is "pancarcinoma" in nature. It is not expressed in most adult tissues with the exception of secretory endometrium (Thor *et al*, 1987) and transitional colonic mucosa (Wolf *et al*, 1989). Limited reactivity of MAb B72.3 was observed in a few benign lesions of the breast and colon. TAG-72 antigen expression was detected, however, in same fetal tissues (colon, stomach, and esophagus), thus defining TAG-72 as an oncofetal antigen. The pancarcinoma distribution and lack of significant reactivity with normal adult tissues of MAb B72.3 suggest its potential diagnostic and therapeutic utility for human carcinomas. However, one characteristic which has been described for TAG-72 as well as other human antigens, is the heterogeneous expression in human carcinoma lesions (Miller and Heppner 1979; Horand Hand *et al*, 1983; Thor *et al*, 1986). Therefore, tumor cells which do not express or express low level of tumor antigen will most likely escape MAb detection, thereby, limiting the diagnostic and therapeutic effects of the MAb. Consequently, antigen heterogeneity is an important consideration when designing MAb-based clinical protocols. One potential approach of overcoming the limitations of heterogeneous tumor antigen expression is to identify compounds able to selectively enhance the level of tumor antigen expression. To date, the list of such agents includes several organic solvents, cyclic AMP and related analogues, vitamin A and D, DNA intercalating agents (*i.e.*, 5-azacytidine), tumor necrosis factors, transforming growth factors, interleukins, several butyrate compounds, and the human interferons (IFNs) (Friedman *et al*, 1987; Niles *et al*, 1988; Chakrabarty *et al*,

1988; Attallah *et al*, 1979; Hwang *et al*, 1986; Guadagni *et al*, 1989). Recombinant human interferons can induce and/or amplify the expression of class I and class II MHC antigens, Fe receptors (Heron *et al*, 1978; Wallach *et al*, 1982) and tumor antigens on the surface of a wide variety of human cell lines (Greiner *et al*, 1984). Reports have demonstrated that the expression of a 90 kDa CEA-related tumor antigen in WiDR tumor xenografts in athymic mice was increased following *in vivo* administration of interferon-α (A) (IFN-αA) and accompanied by a concomitant enhancement in the targeting of a MAb to human carcinoma lesions (Greiner *et al*, 1987a, b; Guadagni *et al*, 1988). In a recent study, human carcinoma cells isolated from malignant effusions of patients diagnosed with different adenocarcinomas were treated *in vitro* with type I [*i.e.*, IFN-α(A), IFN-β_{ser}] or type II (*i.e.*, IFN-γ). It was observed that both types of interferons enhanced the expression of such human tumor antigens as CEA and TAG-72 (Guadagni *et al*, 1989). In addition, cells isolated from non-malignant ascites and treated *in vitro* with interferon did not express either TAG-72 or CEA before or after treatment. These results indicate that interferon treatment does not induce *de novo* expression of these human tumor antigens. Therefore, it may be possible to selectively enhance tumor antigen expression in a carcinoma lesion while the surrounding normal tissues remain tumor antigen negative.

These experimental investigations have led to the design of a phase 1A clinical trial to determine whether IP IFN-γ administration could upregulate TAG-72 and CEA expression on the surface of human carcinoma cells *in vivo* (Greiner *et al*, 1992). These results indicate the potential clinical usefulness of combining on agent able to alter the antigenic phenotype of a tumor cell population with conjugated MAbs, and the possibility of improving the diagnostic and therapeutic efficacy of the MAbs.

IFN-regulation of human tumor antigens: Pre-clinical *in vitro* and *in vivo* studies

A series of anti-CEA MAbs (designed COL-1 through 15) have been generated using extracts or membrane-enriched fractions of primary or metastatic colon lesions (Muraro *et al*, 1985). Immunohistochemical and RIA analyses demonstrated that COLs-1, -4, -12 are able to recognize distinct CEA epitopes (Kuroki *et al*, 1989). These MAbs, along with MAb B6.2 which recognizes a different antigen of 90 kDa glycoprotein normal crossing reacting antigen (NCA-related to CEA), were used to study tumor antigen regulation by recombinant human IFN-αA. The results of these studies showed that IFN can significantly enhance the level of expression of a variety of human tumor-associated antigens (Greiner *et al*, 1984, 1987a b; Guadagni *et al*, 1988). The addition of 1000 U/ml of IFN-αA to a human colorectal carcinoma cell line (WiDR) can increase the level of binding of MAbs (*i.e.*, COL-4 and B6.2) specific for tumor antigens (from 135 to 280%). Likewise the IFN-αA-administration of 250,000 units daily for 6 days, to athymic mice bearing WiDR grown as subcutaneous tumors resulted in a substantial enhancement of MAb tumor binding. Furthermore, a human melanoma cell line (A375), which did not constitutively express either TAG-72 or CEA, remained negative for the expression of these tumor antigens after either *in vitro* or *in vivo* interferon treatment. The results also established that interferon treatment can enhance the localization of a radioconjugated MAb to a human tumor xenograft grown in athymic mice. Additional studies have shown that IFN-β_{ser} and IFN-γ were also capable of inducing significant alterations in the antigenic phenotype of human tumor cells (Kantor *et al*, 1989; Greiner *et al*, 1990; Guadagni *et al*, 1990a). Moreover, those findings revealed that the increase in CEA expression associated with treatment with IFN-γ is accompanied by an increase in the level of CEA-related mRNA transcripts (Kantor *et al*, 1989; Guadagni *et al*,

1990b). These findings suggest that IFN-γ treatment regulates CEA expression at transcriptional and/or posttranscriptional sites which is reminiscent of previous observations for IFN-γ regulation of class II HLA antigens (Rosa and Fellous, 1988).

B72.3 which reacts with the TAG-72 antigen has been successfully used in clinical trials for the localization of metastatic colorectal and ovarian carcinoma lesions (Colcher *et al*, 1987 a, b; Lastoria *et al*, 1988; Schlom *et al*, 1989, 1991). In addition, immunohistochemical studies have reported that the expression of the TAG-72 antigen in most human carcinomas is highly heterogeneous (Thor *et al*, 1986). Therefore, we were interested to determine whether treatment of human tumor cell types which constitutively express TAG-72 resulted in an increased amount of the antigen expressed. As stated previously, this could, therefore, be exploited to increase MAb binding both *in vitro* and *in vivo*. Nevertheless, the *in vitro* study of the TAG-72 regulation by the interferons has been difficult because few established human cell lines constitutively express this antigen (Horan Hand *et al*, 1985; Thor *et al*, 1986). In fact the only adherent human cell lines that constitutively express TAG-72 are the MCF-7 (breast carcinoma) and the highly differentiated LS174T colorectal cell line (Greiner *et al*, 1987a, b). Studies have shown, however, that the LS174T cells are unresponsive to the ability of either type I or type II IFNs to regulate normal HLA as well as tumor antigen expression (Guadagni *et al*, 1990b).

Previous studies demonstrated that *in vitro* treatment of MCF-7 cells with recombinant interferon can enhance TAG-72 expression (Greiner *et al*, 1984), and other reports (Johnston *et al*, 1985) have shown that a high percentage of human malignant effusions constitutively express TAG-72. Therefore, we decided to design a study in human cells isolated from a variety of human serous effusions of patients diagnosed with adenocarcinoma or other non-malignant diseases. The antigen phenotype of each cell population was determined by the binding of various MAbs which recognize normal membrane antigens (*i.e.*, HLA, cytokeratins) as well as those which react with human tumor antigens (TAG-72, CEA). The cells were subsequently treated *in vitro* with IFN-α(A), IFN-β_{ser} or IFN-γ and the changes in the level of expression of the various cell surface antigens analyzed by various immunological assays (*i.e.*, radio-immunoassay, flow cytometry, immunocytochemistry). In particular this study represented an opportunity to investigate the ability of recombinant interferons to regulate TAG-72 expression *in vitro* on different human carcinoma cell populations. A large variety of serous effusions of patients diagnosed with adenocarcinoma (*i.e.*, ovarian, pancreas, breast, unknown) non-epithelial tumors (*i.e.*, melanoma, lymphoma, sarcoma) and benign diseases (*i.e.*, reactive mesothelium) were analyzed for the constitutive expression of TAG-72 and CEA as well as changes in the level of antigen expression (*i.e.*, MAb binding) as a results of interferon treatment (Guadagni *et al.*, 1989). Constitutive TAG-72 expression was found in 35 of the 43 effusions cytologically diagnosed as adenocarcinoma. Treatment with either IFN-α(A), IFN-β_{ser} or IFN-γ increased the level of binding of B72.3 at least 50% above that measured on the untreated, control cells in 27 of 35 (77.1%) of the samples. Furthermore, of the 22 effusions that were CEA positive, interferon treatment increased the binding of the anti-CEA MAb COL-4 in 59.1% of the cases. Of the 43 effusions from primary adenocarcinoma patients studied, 42 (97.7%) constitutively expressed either TAG-72 or CEA on their cell surface, and interferon treatment increased B72.3 and/or COL-4 cell surface binding in 36 of the 42 antigen-positive cases (85.7%). Moreover, those effusions obtained from patients with non-epithelial tumors or benign diseases, failed to express either tumor antigen (TAG-72 or CEA) before or after interferon treatment.

Regulation of antigen expression in patients treated with interferon

The IFN-mediated increase of tumor antigen expression on the surface of malignant cells has also been demonstrated *in vivo*. Several *in vivo* studies have demonstrated that the ability of IFN to augment tumor antigen expression can result in an increased amount of MAb localized to human tumor xenografts (Greiner *et al*, 1987; Guadagni *et al*, 1988), which, in some cases, mediated suppression of tumor growth. These experimental findings have raised the issue of whether IFN could enhance human antigen expression in a clinical setting, thereby improving MAb directed diagnosis and/or therapy of carcinoma lesions. A study in five patients with melanoma administered with IFN-alpha, showed the increased tumor-distribution of a radiolabeled antimelanoma MAb (MAb 96.5) (Rosenblum *et al*, 1988). Tissue samples, however, were not analyzed for changes in tumor antigen expression after IFN-treatment. In a recent study, intraperitoneal (IP) administration of IFN-γ dramatically increased TAG-72 and CEA expression on the surface of malignant ascites cells (Greiner *et al*, 1992). Eight patients diagnosed with ovarian (n=6) or gastrointestinal (n=2) carcinoma with secondary malignant ascites were intraperitoneally administered with escalating weekly doses (0.1-100 MU) of recombinant human IFN-γ for 8 weeks. Prior to the initial IP injection of 0.1 MU an ascites sample was removed, and constitutive TAG-72 and CEA expression were determined on the isolated tumor cells using immunocytochemical staining and flow cytometry. The percentage of constitutively TAG-72-positive carcinoma cells isolated from seven of the eight patients ranged from 10-70%. Cells isolated from one patient did not express TAG-72. IFN-γ administered IP dramatically increased TAG-72 (as measured by binding of anti-TAG-72 MAb B72.3) expression on the surface of carcinoma cells of all patients which constitutively expressed the antigen before treatment. In some cases, IFN-γ treatment increased the percentage of MAB B72.3-reactive tumor cells from 10% to greater than 90% as seen using flow cytometry, and those changes were further corroborated by similar increases in MAb staining intensity observed with immunocytochemical analysis. In addition, ascites-derived tumor cells from two patients with gastrointestinal carcinoma also expressed enhanced CEA levels after IFN-γ treatment. The increased TAG-72 and CEA expression were observed at low IFN-γ doses (*i.e.*, 0.1 to 1.0 MU), which were well tolerated by all patients. This study together with those from experimental studies suggest the possible use of IFN with MAbs in immunodiagnostic and/or therapeutic settings. The increased expression of human tumor antigens such as TAG-72 and CEA by IFN-γ indicate synthesis of new antigen, which is demonstrated as an increase in the percentage of tumor cells expressing the antigens and/or an increased amount of antigen expressed per cell. These changes result in a more homogeneous tumor antigen-positive population. It has previously been shown that the antigen-density on tumor cells is believed to be important for the optimal effectiveness of a MAb to elicit antibody-dependent monocyte cytotoxicity (Herlyn *et al*, 1985). Although no such requirement has been demonstrated for a radionuclide-conjugated MAb, it is assumed that the higher number of binding sites per tumor cell, the better. Thus, the ability to augment human tumor antigen expression should improve the diagnostic and therapeutic use of a MAb. Additional studies are needed to establish the optimal dose and schedule required to maximize increases in antigen expression and to continue to explore the role of IFN-γ or other cytokines as adjuvant for MAb use in immunoscintigraphy and subsequent immunotherapy.

REFERENCES

Attallah AM, Needy CF, Noguchi PD, Ellserg BL (1979) Enhancement of carcinoembryonic antigen expression by interferon. Int J Cancer 24:49-52

Chakrabarty S, Tobon A, Varani J, Brattain MG (1988) Induction of carcinoembryonic antigen secretion and modulation of protein secretion/expression and fibronectin/laminin expression in human colon carcinoma cells by transforming growth factor-β. Cancer Res 48:4059-4064

Colcher D, Carrasquillo JA, Esteban JM, Sugarbaker P, Reynolds JC, Siler K, Bryant G, Larson SM, Schlom J (1987a) Radiolabeled monoclonal antibody B72.3 localization in metastatic lesions of colorectal patients. J Nucl Med 14:251-262

Colcher D, Esteban JR, Carrasquillo JA, Sugarbaker P, Reynolds JC, Bryant G, Larson SM, Schlom J (1987b) Quantitative analysis of selective radiolabeled monoclonal antibody localization in metastatic lesions of colorectal patients. Cancer Res 47:1185-1189

Colcher D, Milenic DE, Ferroni P, Carrasquillo JA, Reynolds JC, Roselli M, Larson SM, Schlom J (1990) *In vivo* fate of monoclonal antibody B72.3 in patients with colorectal cancer. J Nucl Med 31:1133-1137

Friedman J, Seger M, Levinsky H, Allolouf D (1987) Modulation of carcinoembryonic antigen released by HT-29 colon carcinoma cell line in the presence of different agents. Experentia 43:1121-

Greiner JW, Horand Hand P, Noguchi P, Fisher PB, Pestka S, Schlom J (1984) Enhanced expression of surface tumor-associated antigens on human breast and colon tumor cells after recombinant human leukocyte interferon treatment. Cancer Res 44:3208-3214

Greiner JW, Guadagni F, Noguchi P, Pestka S, Colcher D, Fisher PB, Schlom J (1987a) Recombinant interferon enhances monoclonal antibody-targeting of carcinoma lesions *in vivo*. Science 235:895-899

Greiner JW, Horan Hand P, Colcher D, Weeks M, Thor A, Noguchi P, Pestka S, Schlom J (1987b) Modulation of human tumor antigen expression. J Lab Clin Med 109:244-263

Greiner JW, Guadagni F, Horand Hand P, Pestka S, Noguchi P, Fisher PB and schlom J (1990) Augmantation of tumor-associated antigen expression by recombinant human interferons: Enhanced targeting of monoclonla antibodies to carcinomas. In: Goldenberg DM (ed) Cancer Imaging with radiolabeled antibodies, Nijhoff Norwell, Boston, Mass, USA, p 421-435

Greiner JW, Guadagni F, Goldstein D, Smalley RV, Borden EC, Simpson JF, Molinolo A, Schlom J (1992) Intraperitoneal administration of interferon-gamma to carcinoma patients enhances expression of tumor-associated glycoprotein-72 and carcinoembryonic antigen on malignant ascites cells. J Clin Oncol 10:735-746

Guadagni F, Schlom J, Pothen S, Pestka S and Greiner JW (1988) Parameters involved in the enhancement of monoclonal antibody targeting *in vivo* with recombinant interferon. Cancer Immunol Immunother 26:222-230

Guadagni F, Schlom J, Johnson WW, Szpak CA, Goldstein D, Smalley R, Simpson JF, Borden EC, Pestka S, Greiner JW (1989) Selective interferon-induced enhancement of tumor-associated antigens on a spectrum of freshly isolated human adenocarcinoma cells. J Natl Cancer Inst 81:502-512

Guadagni F, Kantor J, Schlom J and Greiner JW (1990a) Regulation of tumor antigen expression by recombinant interferons. In Fisher PB (ed) Mechanism of Differentiation, CRC Press, Boca Raton, FLA, USA, p 35-51

Guadagni F, Witt PL, Robbins PF, Schlom J, Greiner JW (1990b) Regulation of carcino embryonic antigen in different human colorectal tumor cells by interferon-γ. Cancer Res 50:6248-6255

Guadagni F, Tortora G, Clair T, Cho-Chung YS, Schlom J, Greiner JW (1991) Regulation of carcinoembryonic antigen expression on human colorectal tumor cells by a site-selective cyclic Amp analogue. Int J Cancer 48:413-422

Herlyn J, Powe J, Ross AH (1985) Inhibition of human tumor growth by IgG2a monoclonal antibodies with antibody density on tumor cells. J Immunol 134:1300-1304

Heron I, Hokland M, Berg K (1978) Enhanced expression of β_2-microglobulin and HLA antigens on human lymphoid cells by interferon. Proc Natl Acad Sci 75:6215-6219

Horand Hand P, Nuti M, Colcher D, Schlom J (1983) Definition of antigen heterogeneity and modulation among human mammary carcinoma cell populations using monoclonal antibodies to tumor-associated antigens. Cancer Res 43:728-735

Horan Hand P, Colcher D, Salomon D, Ridge JA, Noguchi P and Schlom J (1985) Influence of spatial configuration of carcinoma cell populations on the expression of a tumor associated glycoprotein. Cancer Res 45:833-840

Hwang WI, Sack TL, Kim YS (1986) Effects of cyclic adenosine 3'-5'-monophosphate upon glycoprotein and carcinoembryonic antigen synthesis and release by human colon cancer cells. Cancer Res 46:3371-3374

Johnston WW, Szpak CA, Lottich SC, Thor A and Schlom (1985) Use a monoclonal antibody (B72.3) as a immunocytochemical adjunct to diagnosis of adenocarcinoma in human effusions. Cancer Res 45:1894-1900

Kantor J, Tran R, Greiner J, Pestka S, Fisher PB, Shiveley JE and Schlom J (1989) Modulation of carcinoembryonic messenger RNA levels in human colon carcinoma cells by recombinant human γ-interferon. Cancer Res 49, 2651-2655

Kohler G, Milstein C (1975) Continuous cultures of fused cells secreting antibody of predefined specificity. Nature (Lond.) 256:295-297

Kuroki M, Greiner JW, Simpson JF, Primus FJ, Guadagni F and Schlom J (1989) Serological mapping and biochemical characterization of the carcinoembryonic antigen epitope using fourteen distinct monoclonal antibodies. Int J Cancer 44:208-218

Lastoria S, D'Amico P, Mansi L, Giordano GG, Rossiello R, Schlom J, Pace E, Panza N, Pacilio G and Salvatore M (1988) A prospective imaging study of 131I-B72.3 monoclonal antibody in patients with epithelial ovarian cancer: preliminary report. Nucl Med Commun 9:347-356

Miller PR, Heppner GH (1979) Immunological heterogeneity of tumor cell subpopulations from a single mouse mammary tumor. J Natl Cancer Inst 63:1457-1644

Muraro R, Wunderlich D, Thor A, Lundy J, Noguchi P, Cunningham R and Schlom J (1985) Definition of monoclonal antibodies of a repertoire of epitopes on carcinoembryonic antigen differentially expressed in human colon carcinomas versus adult tumors. Cancer Res 45: 5769-5780

Niles RM, Wihelm SA, Thomas P, Zamcheck N (1988) The effect of sodium butyrate and retinoid acid on growth and CEA production in a series of human colorectal cell lines representing different states of differentiation. Cancer Invest 6:39-45

Rosa FM, Fellous M (1988) Regulation of HLA-DR gene by IFN-γ. Transcriptional and post-transcriptional control. J Immunol 140:1660-1664

Rosenblum MG, Lamki LM, Murray JL, Carlo DJ, and Gutterman J (1988) Interferon-induced changes in pharmacokinetics and tumor uptake of 111In-labeled antimelanoma antibody 96.5 in melanoma patients. J Natl Cancer Inst 80:160-166

Schlom J, Colcher D, Roselli M, Carrasquillo JA, Reynolds JC, Larson SM, Sugarbaker P, Tuttle SE ans Martin EW (1989) Tumor targeting with monoclonal antibody B72.3. Nucl Med Biol 16:137-142

Schlom J, Milenic D, Roselli M, Colcher D, Bird R, Johnson S, Hardman KD, Guadagni F, Greiner JW (1991) New concepts in monoclonal antibody based radio-immunodiagnosis and radioimmunotherapy of carcinoma. Nucl Med Biol 18:425-435

Thor A, Gorstein F, Ohuchi N, Szpak CA, Johnston WW, Schlom J (1986) Tumor-associated glycoprotein (TAG-72) in ovarian carcinomas defined by monoclonal antibody B72.3. J Natl Cancer Inst 76 995-1006

Thor A, Viglione MJ, Muraro R, Ohuchi N, Schlom J (1987) Monoclonal antibody B72.3 reactivity with human endometrium: a study of normal and malignant tissues. Int J Gynecol Pathol 6: 235-247

Wallach D, Fellous M and Revel M (1982) Preferential effect of gamma interferon on the synthesis of HLA antigens and their MRNAS. Nature 299, 833-836

Wolf BC, D'Emilia JC and Salem RR (1989) Detection of the tumor-associated glycoprotein antigen (TAG-72) in premalignant lesions of the colon. J Natl Cancer Inst 81:1913-1917

EFFECTS OF TUMOR NECROSIS FACTOR-ALPHA ON GROWTH AND DOXORUBICIN SENSITIVITY OF MULTIDRUG RESISTANT TUMOR CELL LINES.

M. Crescimanno, N. Borsellino, V. Leonardi, L. Rausa, N. D'Alessandro*.

Istituto di Farmacologia, Facolta' di Medicina e Chirurgia, Universita' di Palermo, Policlinico "P. Giaccone", 90127 Palermo, Italy. *Istituto di Farmacologia, Facolta' di Medicina e Chirurgia, Universita' di Messina, Piazza XX Settembre 4, 98122 Messina, Italy.

Biological agents might offer various therapeutic opportunities in the treatment of cancer, including a direct and/or host- mediated antiproliferative effect as well as the possibility to favourably modulate tumor sensitivity to antineoplastic drugs (Alexander et al., 1987; Kikuchi et al., 1992; Wadler and Schwartz, 1990). However, information on their activity on chemoresistant tumors is still scanty (Billi et al., 1991; Bonavida et al., 1989; D'Alessandro, 1993; Fruehauf et al., 1991; Liddill et al., 1988; Mihich and Ehrke, 1991). Here we have focused on tumor necrosis-alpha (TNF-α) and studied its in vitro effects on the growth of two tumor cell lines, the mouse B16 melanoma and Friend erythroleukemia, both as doxorubicin (DXR)-sensitive (B16, FLC) and -resistant (B16/DXR, FLC/DXR) variants; the resistant lines were endowed with "typical" multidrug resistance (MDR), including cross-resistance to vincristine and the overexpression of P-glycoprotein, detected by immunocytochemistry or immunoblotting (Crescimanno et al., 1991; Schisselbauer et al., 1989). In addition, since in some tumors TNF-α may act through oxidative stress (Zimmerman et al., 1989; Wong and Goeddel, 1988) with greater efficacy in cells depleted of their glutathione content (Zimmerman et al., 1989), we studied the antiproliferative effects of the cytokine in combination with glutathione-depleting concentrations of buthionine sulfoximine (BSO). Finally, experiments were performed to ascertain whether co-treatment with TNF-α could modify the sensitivity to DXR of the drug-sensitive and -resistant cell lines.

SENSITIVITY OF THE CELL LINES TO THE ANTIPROLIFERATIVE EFFECTS OF TNF-α.

A 72-hour exposure to TNF-α (recombinant mouse TNF-α was obtained from Genzyme, Cambridge, MA, USA) inhibited the growth of B16 and B16/DXR with a slightly sloping dose-response curve (Fig. 1, A). The MDR variant, B16/DXR, was slightly more sensitive

NATO ASI Series, Vol. H 75
Cancer Therapy
Edited by N. D'Alessandro, E. Mihich, L. Rausa, H. Tapiero, and T. R. Tritton

than B16. In the case of Friend leukemia, preferential sensitivity to TNF-α of the chemoresistant variant was observed again and it was more pronounced (Fig. 1,B and 2,B). After 3 days, 1000 U/ml of the cytokine inhibited the growth of FLC/DXR by 52%; this effect was not increased substantially by escalating the doses. Instead, FLC was almost completely unresponsive to TNF-α in the whole range of the concentrations tested (Fig.1,B) With a longer exposure to TNF-α (120 hours), an antiproliferative effect of the cytokine appeared also in FLC, but still to a lesser degree than in FLC/DXR (Fig. 2, B).

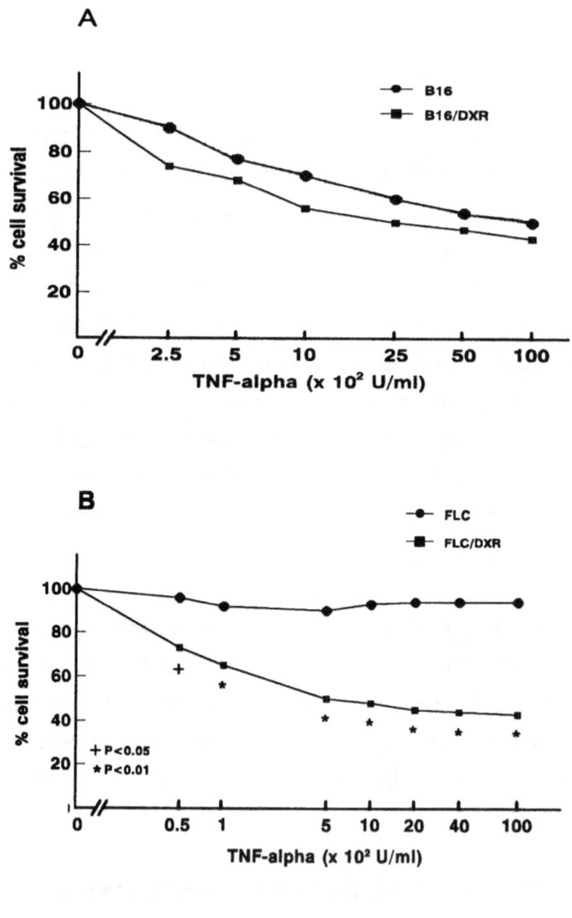

Figure 1. Effect of TNF-α on the growth of B16 and B16/DXR cells (A), FLC and FLC/DXR cells (B) exposed for 72 hours to the biological agent. Results presented are the mean of three independent experiments performed in triplicate. SD were regularly ≤ 10% of the mean.

EFFECT OF BSO.

In these experiments the cells were seeded in the presence of a fixed concentration of BSO; 24 hours later, various concentrations of TNF-α were added and the cells were counted after a further 48 hours (also 120 hours in the case of FLC and FLC/DXR).

The basal glutathione content (determined according to Griffith, 1979) was 35.4 ± 2.6 nmoles/mg protein in B16 and 47.8 ± 1.1 in B16/DXR. BSO 50 μM for 3 days reduced the glutathione level to about 2% of the basal content both in B16 and B16/DXR. After 3 days this concentration of BSO depressed the growth of B16 by 20% and that of B16/DXR by 32% (not shown); in plotting the combination data (Fig 2, A), these effects of BSO alone were compensated by making these equal to 100%. As illustrated in Fig. 2, A, the effects of BSO 50 μM plus TNF-α on the growth of B16 and B16/DXR were slightly but consistently synergistic.

The basal glutathione content was 8.8 ± 0.5 nmoles/mg protein in FLC and 23.6 ± 0.9 in FLC/DXR; BSO 100 μM for 3 days lowered this level to 23.8% in FLC and to 8.0% in FLC/DXR. The combination of BSO 100 μM (for 72 hours) and various concentrations of TNF-α (for 48 hours) produced only additive effects on the growth of both the cell lines (not shown); however, when FLC and FLC/DXR were exposed to BSO 100 μM plus TNF-α for a more prolonged time (Fig. 2, B), a significant synergism of the combination was observed in FLC/DXR. It must be added that BSO 100 μM alone for 6 days did not modify the growth of FLC or FLC/DXR.

EFFECT OF TNF-α ON THE SENSITIVITY TO DXR OF THE CELL LINES.

In these experiments, the cells were seeded in the presence of a slightly cytotoxic concentration of TNF-α; 24 hours after DXR was added and the cells were counted after an additional 72 hours.

For B16 and B16/DXR, the dose of TNF-α used was 500 U/ml which alone lowered the growth of B16 by 22% and that of B16/DXR by 30% (not shown); in plotting the combination data of Fig. 3, A, these effects of TNF-α alone were compensated by making these equal to 100%. It can be seen (Fig. 3, A) that TNF-α plus DXR produced additive effects on the growth of B16 and synergistic ones on that of B16/DXR.

For FLC and FLC/DXR the dose of TNF-α used was 200 U/ml which alone reduced the growth of FLC by 11% and that of FLC/DXR by 36% (not shown); in plotting the combination data of Fig. 3, B, these effects of TNF-α alone were again compensated by making these equal to 100%. The combination of TNF-α and DXR produced only additive effects on the growth of FLC and FLC/DXR (Fig. 3, B).

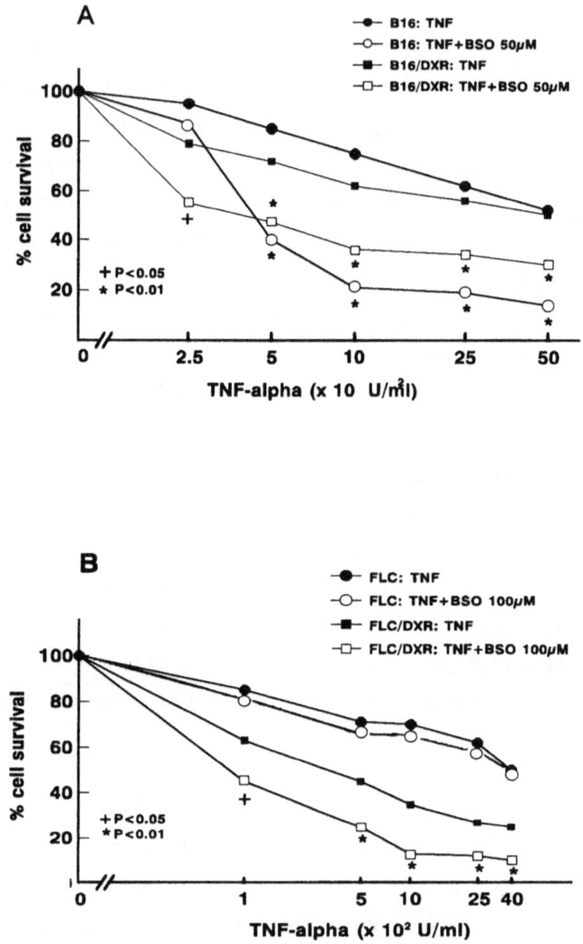

Figure 2. Effect of TNF-α ± BSO on the growth of B16 and B16/DXR cells (A), FLC and FLC/DXR cells (B). The cells were seeded in the presence of BSO or not; 24 hours after, various doses of TNF-α were added and the cells were counted (through the microscope with trypan bleu exclusion) after a further 48 hours for B16 and B16/DXR, and 120 hours for FLC and FLC/DXR. Combined TNF-α/BSO data are normalized to account for the cell survival inhibitory effect of BSO alone (see text). Results presented are the mean of three independent experiments performed in triplicate. SD were regularly ≤ 10% of the mean.

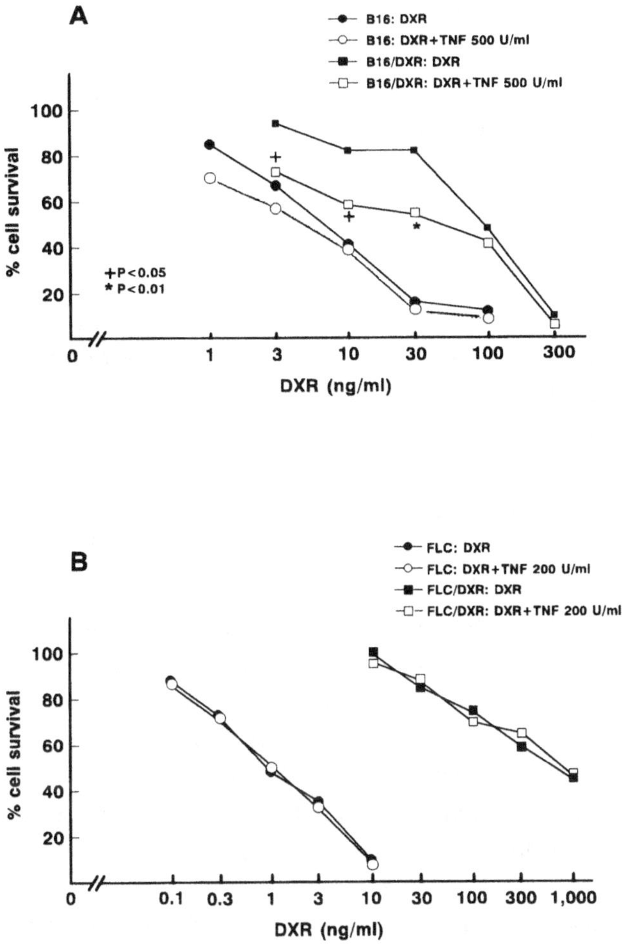

Figure 3. Effects of DXR ± TNF-α on the growth of B16 and B16/DXR cells (A), FLC and FLC/DXR cells (B). The cells were seeded in the presence of TNF-α or not. After 24 hours, various doses of DXR were added and the cells were counted (through the microscope with trypan bleu exclusion) after a further 72 hours. Combined DXR/TNF-α data are normalized to account for the cell survival inhibitory effect of TNF-α alone (see text). Results presented are the mean of three independent experiments performed in triplicate. SD were regularly ≤ 10% of the mean.

CONCLUSIONS.

To sum up, we studied the effects of TNF-α on the in vitro growth of two cell lines, the mouse B16 melanoma and Friend erythroleukemia, as parental and DXR-resistant (MDR) variants. All these tumours showed moderate sensitivity to the cytokine with a dose-response relationship but it is interesting to note that the MDR cell lines (especially FLC/DXR) responded better than their sensitive counterparts. Favourable changes in the sensitivity to the antitumor effects of TNF-α have already been described in other MDR variants (Liddill et al., 1988); however, the contrary has been observed by other authors (Fruehauf et al., 1991). In any case, our results and those of others suggest that the acquisition of a MDR phenotype may alter the response to the biological agent, a notion that could perhaps be relevant in the clinical context.

Clearly, many factors have been put forth which may be responsible for the tumor cell sensitivity to TNF-α; for example, they could involve changes in the specific binding of the cytokine to the membrane of the target cells and in its internalization as well as other processes including those leading to the fragmentation and repair of DNA following the treatment (Fruehauf et al., 1991; Nio et al., 1991). It has also been suggested that an enhanced free radical generation and, counteracted to it, the efficiency of the antioxidant activities such as manganous superoxide dismutase (Wong and Goeddel, 1988) or glutathione (Zimmerman et al., 1989) may be involved in the antitumor effects of the biological agent. As a matter of fact, in our experiments glutathione-depleting concentrations of BSO synergized with the growth inhibitory effects of TNF-α in those cell lines which had had a better response to the cytokine alone, i.e. B16, B16/DXR and FLC/DXR; this seems to confirm that oxidative stress plays a role in the antitumor effects of TNF-α.

Finally, the possibility that TNF-α might increase cell responsiveness to DXR was taken into account. A synergism was encountered in the MDR variant of B16 melanoma, but not in the other tumors. We are not currently aware of and are trying to identify the causes that may account for the successful interaction between TNF-α and DXR in this particular cell line. Other authors have described the favourable influences of TNF-α on the activity of various cytotoxic drugs, including DXR, in drug-sensitive and -resistant cells (Alexander et al., 1987; Billi et al., 1991; Bonavida et al., 1989; Fruehauf et al., 1991; Kikuchi et al., 1992). In the case of topoisomerase II- targeted drugs, such as DXR, TNF-α might potentiate their activity on this enzyme; the order in which the components of the combination are added might be critical, with better interactive effects when the exposure to the cytotoxic drugs precedes that to TNF-α (Alexander et al., 1987; Kikuchi et al., 1992). We are now verifying this possibility.

Of course, herein we have taken only one of the possible aspects of the antitumor activity of TNF-α into account, i.e. that of its direct effects at the cell level. Certainly, it is worth carrying out further studies in more complex, especially in vivo models, in order to assess fully the therapeutic activity of TNF-α as well as of other biological agents, on drug - sensitive and -resistant cancer.

ACKNOWLEDGEMENTS. This work was partially supported by AIRC and by MURST 60%.

REFERENCES

Alexander RB, Nelson WG, Coffey DS (1987) Synergistic enhancement by tumor necrosis factor of in vitro cytotoxicity from chemotherapeutic drugs targeted at topoisomerase II. Cancer Res 47: 2403-2406.

Billi G, Venturini M, Mazzoni A, Addamo GF, Russo P (1991) Reversione di at-MDR da parte del TNF in una linea cellulare di carcinoma ovarico. Tumori 77 (supplement 3): 151.

Bonavida B, Tsuchitani T, Safrit J, Zighelboim J (1989) Mechanisms of target cell lysis by cytototoxic cells, factors and drugs. In: Tapiero H, Robert J, Lampidis TJ (eds) Anticancer Drugs, Johm Libbey Eurotext London Paris, pp. 113- 125.

Crescimanno M., Armata MG, Florena AM, Leonardi V, Rausa L, D'Alessandro N (1991) Antioxidant defenses in a B16 melanoma line resistant to doxorubicin: an in vivo study. Anti-Cancer Drugs 2: 481-486.

D'Alessandro N (1993) Immunological aspects of tumour drug-resistance. In press on: Current topics in pharmacology, Research Trends, Trivandrum, India.

Fruehauf JP, Mimnaugh EG, Sinha BK (1991) Doxorubicin induced cross-resistance to tumor necrosis factor related to altered TNF processing. Proc Am Assoc Cancer Res 32: 269.

Griffith OW (1979) Determination of glutathione and glutathione disulfide using glutathione reductase and 2-vinyl-piridine. Annal Biochem 106: 207-214.

Liddill JD, Dorr RT, Dalton WS, Salmon SE (1988) Correlation of multidrug resistance to recombinant necrosis factor sensitivity based on lysosomal enzyme activity. Proc Am Assoc Clin Oncol 7: 50.

Kikuchi A, Holan V, Minovada J (1992) Effects of tumor necrosis factor alpha, interferon alpha and interferon gamma on non-lymphoid leukemia cell lines: growth inhibition, differentiation induction and drug sensitivity modulation. Cancer Immunol Immunother 35: 257-263.

Mihich E, Ehrke MJ (1991) Immunomodulation by anticancer drugs. In: De Vita VT, Hellman S, Rosenberg SA (eds) Biologic therapy of cancer, JB Lippincott Company Philadelphia, pp. 776-786.

Nio Y, Zighelboim J, Bonavida B (1991) Effect of recombinant tumor necrosis factor on fragmentation and repair of deoxyribonucleic acid of sensitive and resistant human leukemia cell lines. J Cell Pharmacol 2: 32-40.

Schisselbauer JC, Crescimanno M, D'Alessandro N, Clapper M, Toulmond S, Tapiero H, Tew KD (1989) Glutathione, glutathione-S-transferases and related redox enzymes in adriamycin- resistant cell lines with a multidrug resistant phenotype. Cancer Commun 1: 133-139.

Zimmerman RJ, Marafino BJ, Chan A, Landre P, Winkelhake JL (1989) The role of oxidant injury in tumor cell sensitivity to recombinant human tumor necrosis factor in vivo. J Immunol 142: 1405-1409.

Wadler S, Schwartz EL (1990) Antineoplastic activity of the combination of interferon and cytotoxic agents against experimental and human malignancies: a review. Cancer Res 50: 3473-3486.

Wong GHW, Goeddel DV (1988) Induction of manganous superoxide dismutase by tumor necrosis factor: possible protective mechanism. Science 242: 941-944.

ACTIVATION OF MACROPHAGES BY TREATMENT OF RAT PERITONEAL CELLS WITH PHOTOFRIN II AND He-Ne LASER

V.F.Dima, M.Ionescu, V.Vasiliu and S.V.Dima
Department of Experimental Pathology
Cantacuzino-Institute and Institute of Atomic Physics
Bucharest 70100 - Romania

Some biological properties of macrophages activated with C.parvum, MurIFN-gamma and Photodynamic therapy (PDT) are studied. Results obtained indicated that: (i) macrophages activated with the three immunopotentiators had an intense phagocytary activity; (ii) association of B and T peritoneal lymphocytes treated with PDT to macrophages increased their biological properties; (iii) cytotoxic activity of PDT-AMϕ varied between 31.6 and 78.6% function of the tumor target cells; (iv) the three sets of macrophages exerted cytostatic effects upon human and mouse leukemia cells.
Summing up, macrophage immunopotentiations with the photodynamic therapy has proved to be a useful method for cell biology as well as for adoptive immunotherapy in various cancer forms.

INTRODUCTION

Biological activities (bactericidal, cytotoxic and cytostatic) of macrophages were stimulated by a variety of agents, such as: C.parvum, interferon, phorbol myristate acetate (PMA), LPS and the late complement components C5b-9 (Steubing *et al*, 1991). During the course of the activation of macrophages, reduction of oxygen produces toxic agents such as superoxyde anion, hydrogen peroxide, the hydroxyl radical and singlet molecular oxygen. Recently, it has been proved that activated murine macrophages can induce regression of murine tumors and pulmonary metastases (Chokri *et al*, 1989). Similarly, human macrophages differentiated from blood monocytes and activated by interferon have been used in the adoptive immunotherapy (Bartholeyns *et al*, 1991; Andreesen *and* Hannemann, 1991). The paper reports on the biological functions of macrophages

NATO ASI Series, Vol. H 75
Cancer Therapy
Edited by N. D'Alessandro, E. Mihich, L. Rausa,
H. Tapiero, and T. R. Tritton

activated by the photodynamic treatment as compared to C.parvum and interferon.

MATERIALS AND METHODS

Activation of macrophages, was achieved by 3 methods: (1) "in vivo" after i.p. innoculation of rat with C.parvum (20 mg/kg) and prelevation of peritoneal exudate after 7 days (C.parvum-AMϕ); (2) monocytes isolated from rat blood by the method of Bartholeyns *et al,* 1991) and cultivated in RPMI 1640 ± 10% FCS for 6 days at 37°C and MurIFN - gamma 500 IU/ml activated for 18 h at 37°C (IFN-AMϕ) as compared to (3) photodynamic therapy of rat peritoneal macrophages sensitized with different doses of Photofrin II (0.01-10.0 μg/ml/10^6 cells) and irradiated with He-Ne laser (632.8 nm; 10 mW) at dose fluences varying between 1.5 and 18 kJ/m^2 (PDT-AMϕ). The macrophage activation degree was estimated by means of the following methods.

PHAGOCYTARY PROCEDURE

Enterotoxigenic E.coli-H10407 strain, serotype 078:H11 was cultivated and labelled with ^{3}H-thymidine according to the method described by Dima *et al* (1990). Equal volumes of microbial cell suspension (1x10^7 cell/ml-equivalent to 22.835 cpm) and activated macrophages (1x10^6 cells/ml) were incubated 3 h at 37°C. The radioactive values were read on the Beckman spectrometer and expressed in percentages against the values found in the control.

CYTOTOXICITY ASSAY

Effector cells cytotoxicity was assayed by ^{3}H-Uridine method as described by Nishimura *et al* (1986). Target cell lysis was calculated by the following equation: % cytotoxicity = 1-(cpm in culture of effector and target cells/cpm culture of target cells alone) x 100.

CYTOSTATIC ACTIVITY

Normal and activated macrophages by C.parvum, MurIFN-gamma and

PDT were tested for cytostatic activity by the method described by Tsuchiya *et al* (1983). Cytostatic activity was expressed as percentage of the inhibition of ^{3}H-thymidine incorporation into leukaemia cells. The following leukaemia cell lines were used: K562, MOLT-4, YAC-1 and RL-1.

ELECTRON MICROSCOPY

After 3 h incubation at 37°C normal and activated macrophages mixed with bacteria were fixed, sectioned and examined at electron microscope (Hitachi-H11) according to the method described by Dima *et al* (1989). *Statistics.* Data were expressed as the arithmetic mean ± the standard error (SE). The statistical significance of differences between groups was calculated by the Kruskal-Willis test.

RESULTS

PHOTODYNAMIC ACTIVATION OF PERITONEAL MACROPHAGES

The results obtained revealed an activation of rat peritoneal macrophages following sensitization with different Photofrin II (0.01-10.0 μg/ml) concentrations and exposure to various He-Ne laser irradiation doses (1.5-18 kJ/m^2).The highest rate of bacterial ingestion (283.6%) was found in macrophages sensitized with 0.5 μg/ml Photofrin II and exposed for 60 sec to He-Ne laser irradiation.Phagocytosis of E.coli bacteria by macrophages activated "in vivo" and "in vitro" with C.parvum (335.0%) and MurIFN-gamma (301.2%) was higher than the values found in PDT activated macrophages. The mixing of rat peritoneal (B and T) lymphocytes treated by PDT with macrophages resulted in the enhancement of their phagocytic capacity after 3 h at 37°C. By contrast, microbial ingestion showed a decline (74.8%) when macrophages were mixed with B and T lymphocytes from animals with Alloxan-induced diabetes mellitus.

CYTOTOXIC ACTIVITY OF ACTIVATED MACROPHAGES

The results presented in **Fig. 1** point out the capacity of PDT to stimulate the cytotoxic activity of macrophages against

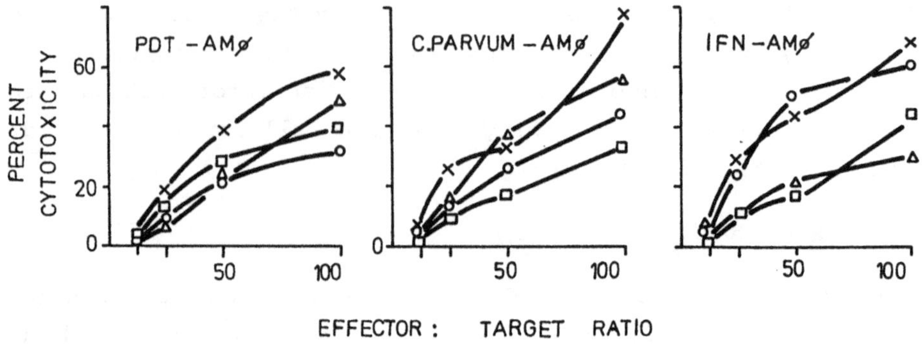

Fig.1 *Tumor cell sensitivity to activated macrophages Cytotoxic activity was measured in 18 h assay at different EC/TC ratio. Spontaneous release of the isotope from target cell was less than 1o.8%. x-x (K562); o-o (MOLT-4); Δ - Δ* (YAC-1) □ - □ *(RL-1).*

target cells (K562, MOLT-4, YAC-1, RL-L). PDT-activated macrophages showed superior values against K562, and MOLT-4 and relatively low values against YAC-1, and RL-1 target cells in comparison with non-stimulated cells (values in the range of 18.5-23.9%). Activation of macrophages by C.parvum and MurIFN-gamma showed higher cytotoxic values against K562, and YAC-1 target cells and medium against MOLT-4 and RL-1. The lysis of target cells was estimated after 18 h at 37°C and gradually increased according to the effector: target ratio (12.5 to 100:1).

CYTOSTATIC EFFECTS

The results obtained on the cytostatic effects of activated macrophages allowed of the following observations: (i) all the three sets of macrophages had a cytostatic activity ranging between 44.6 and 89.6%; (ii) of the three sets the most marked cytostatic activity was that of C.parvum-AMϕ (between 78±9 and 89±8%) against 3 (K562, YAC-1, RL-1) of the 4 leukaemia lines and relatively low against MOLT-4 (53±7%); (iii) PDT-AMϕ had a more significant cytostatic activity against YAC-1 (71±8%) and K562(83±1%) and more reduced against MOLT-4 (55±1%) and RL-1 (44±6%); (iv) the cytostatic activity of IFN-AMϕ was similar with that found for C.parvum-AMϕ and PDT-AMϕ as against the control cells (not-treated).

ELECTRON MICROSCOPY

Electron micrographs of activated macrophages by PDT showed Golgi apparatus, mitochondria and lysosomes in cytoplasm **(Fig. 2)**. **Fig. 3** illustrates (after 3 h at 37°C) the elements in phagocytic process: *(a)* phagosomes with ingested germs; *(b)* phagocytic vacuoles; *(c)* presence of amorphous materials in phagosomes; *(d)* the presence of phagocytized germs but without lytic lesions of bacterial wall and cytoplasm which were observed mostly in C.parvum-activated macrophages **(Fig. 4)**. In macrophages sensitized with high Photofrin II doses (3-10 μg/ml) and prolonged exposure to He-Ne laser irradiation (5-18 kJ/m^2) the following morphological modifications were noticed: (i) partial preservation of nuclei; (ii) cytoplasm vacuolization and (iii) lesions of subcellular organites **(Fig. 5)**.

In the present paper, we have noticed that activation of rat peritoneal macrophages with photodynamic therapy, stimulated their bactericidal, cytotoxic and cytostatic functions. Association of B and T peritoneal lymphocytes treated with PDT to macrophages increased their activities, whereas addition of lymphocytes from animals with diabetes mellitus produced partial inhibition of macrophage activities (Ilonen *et al*, 1991; Yamamoto *et al*, 1991). The results allowed the following observations: (1) importance of lymphocytes in initiation of macrophage activation (increase in fluidity of lymphocyte membrane lipids after PDT); (2) importance of their functional (metabolic) state and (3) the He-Ne laser emitting red light has a powerful effect on activation of macrophages sensitized with Photofrin II (Yamamoto *et al*, 1991).
Summing up, the biological functions (bactericidal, cytotoxic, cytostatic) of PDT photobioactivated macrophages were as efficient as of those immunopotentiated by C.parvum and interferon.

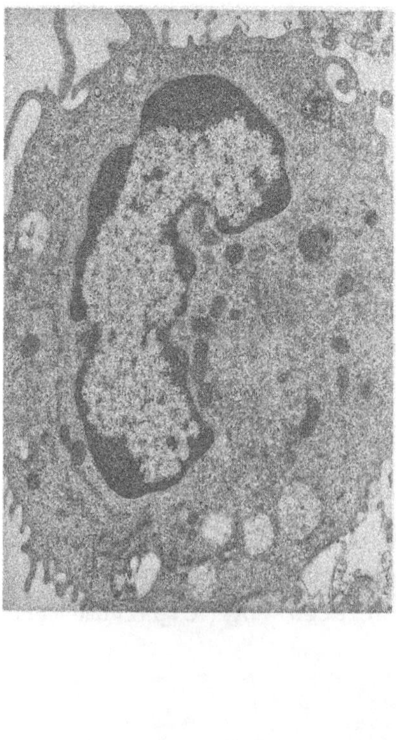

Figure 2

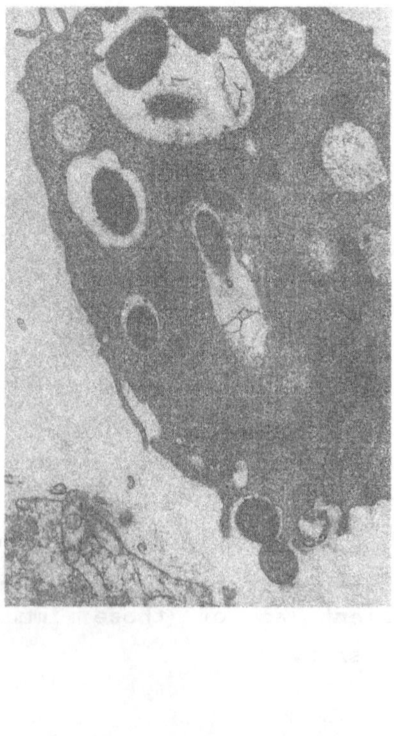

Figure 3

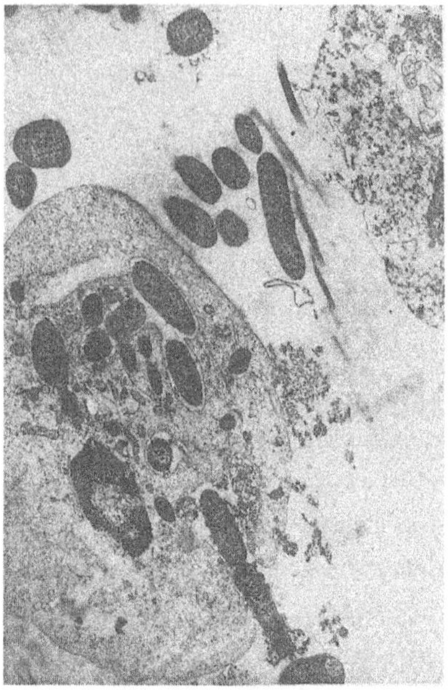

Figure 4

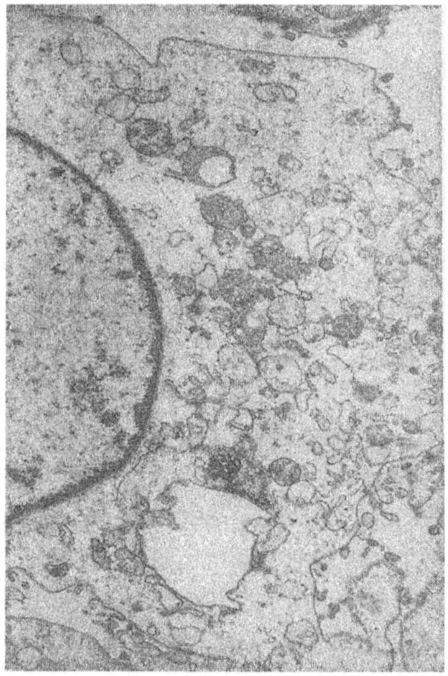

Figure 5

REFERENCES

Andreesen R, Hennemann B (1991) Adoptive immunotherapy with autologous macrophages: Current status and future perspecti - ves. Pathobiology 59:259-265

Bartholeyns J, Lopez M, Andreesen R (1991) Adoptive immunotherapy of solid tumors with activated macrophages: Experimental and clinical results. Anticancer Res. 11:1201-1204

Chokri H, Freudenberg M, Galanos C, Poindron P and Bartholeyns J (1989) Antitumoral effects of lypopolysaccharides, tumor necrosis factor, interferon and activated macrophages: Synergism and tissue distribution. Anticancer Res 9:1185-1190

Dima F V, Petrovici Al and Dima S V (1989) Activation of guinea pig peritoneal macrophages by ribosomal extract of Salmonella typhi strain. Electron microscopy studies. Arch Roum Path exp Microbiol 48:323-340

Dima F V, Petrovici Al and Dima S V (1990) Characterization of colonization factor antigen CFA/I from an enterotoxigenic Escherichia coli strain (0112:H12). Arch Roum Path exp Microbiol 49:297-313

Ilonen J, Surcel M H, Käär L M (1991) Abnormalities within CD4 and CD8 T lymphocytes subsets in type 1 (insulin-dependent) diabetes. Clin exp Immunol 85:278-281

Nishimura T, Togashi Y, Goto M, Yagi H, Uchiyama Y and Hashimoto Y (1986) Augmentation of the therapeutic efficiency of adoptive immunotherapy by in vivo administra - tion of slowly released recombinant interleukin 2. Cancer Immunol Immunother 21:12-18

Steubing W R, Yeturu S, Tuccillo A, Sun H C, Berns W M (1991) Activation of macrophages by Photofrin II during photodyna - mic therapy. J Photochem Photobiol B Biol 10:133-145

Tsuchiya T, Norimura T, Okamoto M (1983) Cell - mediated immunity in host after tumor irradiation. J Radiat Res 24:345-355

Yamamoto N, Homma S, Sery W T, Donoso A L, Hoober K J (1991) Photodynamic immunopotentiation: in vitro activation of macrophages by treatment of mouse peritoneal cells with hematoporphyrin derivative and light. Eur J Cancer 27:467-471

SYNERGIC INTERACTION BETWEEN SIMVASTATIN AND ANTINEOPLASTIC DRUGS ON GLIOMA CELL GROWTH

M.R. Soma, R. Baetta, C. Ferrari*, M.R. de Renzis*, R. Paoletti and R. Fumagalli
Institute of Pharmacological Sciences, University of Milan, Via Balzaretti 9, 20133, Milan Italy. *Centro Istochimica C.N.R., Dipartimento di Biologia Animale, Universita' di Pavia, Pavia, Italy.

INTRODUCTION

The microsomal enzyme 3-hydroxy-3-methylglutaryl coenzyme A (HMG-CoA) reductase produces mevalonate, which is used for the synthesis of cholesterol and several non sterol products in animal cells (Goldstein, Brown, 1990). Inhibition of this enzyme triggers regulatory response that lowers the concentrations of low density lipoproteins (LDL) in blood. A major relationship exists between the cell growth processes and the mevalonate biosynthetic pathway. HMG-CoA reductase activity is high in rapidly growing cells; conversely, when mevalonate synthesis is strongly inhibited cell growth is blocked (Goldstein, Brown, 1990). Thus, it has been suggested that mevalonate may play an important role in cell proliferation. Several studies have demonstrated the need for growing cells of at least two products synthesized from mevalonate in order to proliferate, and the only one yet identified is cholesterol (Habenicht et al.1980; Goldstein, Brown, 1990). Mevalonate is an essential precursor of several important cellular constituents, including: cholesterol, involved in cell membrane maintenance; heme A and ubiquinone, involved in electron transport; dolychol, required for glycoprotein synthesis; and isopentyl adenine, present in some transfer RNAs (Goldstein, Brown, 1990). The most recently discovered products of the HMG-CoA reductase reaction are the prenylated proteins (Glomset et al.1990; Maltese et al.1990). These proteins are attached to the cytoplasm leaflets of cell membranes by virtue of covalently bound isoprenoid groups, either farnesyl (15 carbons) or geranyl-geranyl (20 carbons), both of which are derived from mevalonate. Examples include the growth-controlling ras proteins which are bound to the plasma membrane, and the traffic-directing rab proteins, which are bound to intracellular vesicles. Ras and rab belong to a large family of "small" GTP binding proteins whose activities are controlled by the binding and regulated hydrolysis of GTP (Maltese et

NATO ASI Series, Vol. H 75
Cancer Therapy
Edited by N. D'Alessandro, E. Mihich, L. Rausa, H. Tapiero, and T. R. Tritton

al.1990). A number of highly selective HMG-CoA reductase inhibitors have been developed, the aim being to decrease elevated plasma cholesterol levels (Kathawala, 1991). However, because mevalonic acid is the precursor of numerous metabolites, inhibition of HMG-CoA reductase has the potential to result in pleiotropic effects. This possibility is supported by several "in vivo" and "in vitro" observations showing how some HMG-CoA reductase inhibitors (vastatins) can in fact inhibit cellular proliferation and modulate changes in cell cycle (Maltese, 1984) and cellular morphology (Maltese, Sheridan, 1985). Based on these findings vastatins have received increasing attention as pharmacological tools for controlling abnormal cell growth such as myocyte proliferation under atherogenic conditions or tumor development.

The prognosis of patients with intracranial tumors is poor in spite of the intensive combination therapy. Human brain tumors show considerable resistance to chemotherapy (Paoletti et al.1990). Carmustine (BCNU) is the most commonly used chemotherapeutic agent for brain tumor treatment (Brandes et al.1991). Few reports are available on the therapeutic use of fibroblasts interferon (β type) (Goldstein, Laszlo, 1986). Failure of conventional chemotherapy to produce durable remission against the brain tumors and the presence of important side effects has prompted investigators to seek alternative treatment strategies such as synergistic action between drugs to assure higher efficacy in suppressing tumor proliferation with lower adverse effects. The potential use of vastatins for brain tumor treatment has been suggested since an increased activity of HMG-CoA reductase has been observed in certain such tumors (Maltese, 1983). Furthermore, lovastatin has been shown to inhibit proliferation and induce cell differentiation of rat neuroblastoma cells (Maltese, 1984), as well as to inhibit "in vivo" tumor growth in mice (Maltese et al.1985). We thus investigated the possibility of using this class of drugs, alone or in combination with antineoplastic compounds, to control glioma cell growth in an "in vitro" model (Soma et al.1992).

RESULTS AND DISCUSSION

When exponentially growing human glioma cells were treated with simvastatin, cell growth rate and DNA synthesis were inhibited in a dose-dependent manner with an ID_{50} of 60 nM (Table 1). As expected, simvastatin was also able to inhibit dose-dependently the activity of the HMGCoA reductase enzyme with an EC_{50} of 5 nM.

Table 1. Dose–response effect of simvastatin on glioma cell growth and HMGCoA reductase activity

Simvastatin nM	Cell number	DNA synthesis	HMGCoA R activity
1	92	93	63
10	69	71	36
100	44	44	18
1000	29	27	7

Values are percentage versus control. Cells were incubated with different simvastatin concentrations for 48 hours. The control activity (100%) were 134,325 cells/plate, 388 [^{3}H] thymidine in DNA (cpm/μg protein), and 211 pmole of mevalonate/minute/mg cell prot., for cell number, DNA synthesis and HMGCoA reductase activity, respectively.

In time–course experiments the inhibition of cell growth correlated well with the latent enzyme activity in the cells: the proliferation of glioma cells was inhibited as long as at least 90% of HMGCoA reductase activity was inhibited; the mevalonate requirement for cell growth was not satisfied by exogenous sterols (cells were grown in medium containing the normal supply of cholesterol present in fetal calf serum), suggesting that metabolite(s) other than cholesterol are involved in cell cycle control (Soma et al.1992).

Flow cytometry was utilized to determine whether the non–proliferating cells in the cultures treated with simvastatin were arrested in a particular phase of the cell cycle as reported for lovastatin (Jakobisiak et al.1991). The DNA istograms of cells exposed to increasing concentration of simvastatin showed a marked decrease of cells in the S and G_2+M phases, with an increase in the proportion of cells in the G_1 compartment (Table 2).

Table 2. Dose–response effect of simvastatin on glioma cell cycle

Simvastatin	G_1/G_o	S	G_2 + M
10 nM	103	75	90
5 μM	109	34	46
10 μM	111	16	44

Cells were incubated with different concentration of simvastatin for 24 hours. Values are percentage versus control. The control (100%) were 87.1, 6.1, and 6.8 for G_1/G_o, S, and G_2 + M, respectively.

These results indicate that when HMGCoA reductase enzyme activity is profoundly inhibited, DNA synthesis and cell growth cease and the inhibited cells accumulate in the G_1 phase of the cell cycle. Inhibition of cell growth by simvastatin could be completely reversed incubating cells with mevalonate, demonstrating that the activity of simvastatin is due to its specific inhibitory effect on intracellular mevalonate synthesis (Soma et al.1992).

The effect of simvastatin in combination with two antitumoral agents, BCNU and β interferon, was also explored. Individually, these two drug inhibited cell growth in a concentration–dependent fashion, the ID_{50}'s being 50 nM and 650 U/ml for BCNU and β interferon, respectively (Table 3).

When subliminal concentrations of simvastatin (1 nM), able to inhibit cell growth by 8–10%, were incubated with increasing concentrations of either BCNU or β interferon, a strong inhibition of cell growth could be achieved. The effect is summarized in Table 3. A sensitization factor (SF) was determined by dividing the ID_{50} for the drug alone by the ID_{50} of the drug in the presence of 1 nM simvastatin. Combination of simvastatin with either BCNU or β interferon produced synergistic antigrowth interaction which was 10 and 4.6 times more effective than individual compounds, respectively. Thus, the combination of these antineoplastic drugs with very low doses (in the nanomolar range) of simvastatin permitted the use of several fold less chemotherapeutic agents to achieve the same inhibition cell growth. If simvastatin were used at higher doses (10 and 100 nM) in combination with either BCNU or β interferon, a more pronounced synergistic effect could be achieved.

Table 3. Effect of BCNU and β–interferon alone and in combination with 1 nM simvastatin on human glioma cell growth.

Drug	IC_{50}	Sensitization factor	Type of interaction
BCNU	50 nM	--	
BCNU + Simvastatin	5 nM	10	Synergistic
β–interferon	650 U/ml	--	
β–interferon + Simvastatin	140 U/ml	4.6	Synergistic

Sensitization factors were obtained by dividing the IC_{50}s of the antineoplastic drugs alone and combined with 1 nM simvastatin.

Cells were incubated with drugs for 48 hours.

Interestingly, the addition of mevalonate to cells incubated with 1 nM simvastatin and either BCNU or β–interferon, abolished the synergistic growth inhibition, so that cells proliferated as if they were exposed to BCNU or β–interferon alone. This indicates that mevalonate or one of its products might be directly involved in the synergy observed.

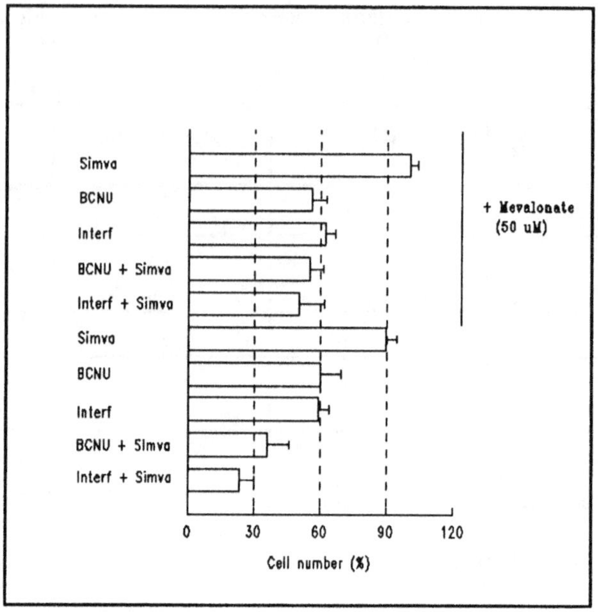

In conclusion, simvastatin, a cholesterol lowering agent, is able to lower the rate of human glioma cell growth in vitro probably through one of its non-sterol derivatives. The drug shows a strong synergistic inhibitory effect on cell proliferation when combined with antineoplastic agents such as BCNU and β-interferon. This could be of potential clinical interest for brain tumor treatment.

Acknowledgements

Supported in part by M.U.R.S.T. (Italian Government) and by C.N.R. Target Project "Biotechnology and Bioinstrumentation"; dual parameter FCM were performed with the instrumentation provided by the "Centro Grandi Strumenti" of Pavia University.

REFERENCES

Brandes A, Soesan M, Fiorentino MV (1991) Medical tretment of high grade malignant gliomas in adults: an overview. Anticancer Res 11:719–728

Glomset J, Gelb M, Farnsworth C (1990) Prenyl proteins in eukaryotic cells: a new type of membrane anchor. Trends Biochem Sci 15:139–142

Goldstein D, Laszlo J (1986) Interferon therapy in cancer: from imaginon to interferon. Cancer Res 46:4315–4329

Goldstein JL, Brown MS (1990) Regulation of the mevalonate pathway. Nature 343:425–430

Habenicht AJR, Glomset JA, Ross R (1980) Relation of cholesterol and mevalonic acid to the cell cycle in smooth muscle and Swiss 3T3 cells stimulated to divide by platelet–derived growth factor. J Biol Chem 255:5134–5140

Jakobisiak M, Bruno S, Skierski JS, Darzynkiewicz Z (1991) Cell cycle–specific effects of lovastatin. Proc Natl Acad Sci 88:3628–3632

Kathawala FG (1991) HMGCoA reductase inhibitors: an exciting development in the treatment of hyperlipoproteinemia. Med Res Rev 11:121–146

Maltese WA (1983) 3–Hydroxy–3–methylglutaryl coenzyme A reductase in human brain tumors. Neurology 33:1294–1299

Maltese WA (1984) Induction of differentiation in murine neuroblastoma cells by mevinolin, a competitive inhibitor of HMGCoA reductase. Biochem Biophys Res Commun 120:454–460

Maltese WA, Defendini R, Green RA, Sheridan KM, Donley DK (1985) Suppression of murine neuroblastoma growth in vivo by mevinolin, a competitive inhibitor of 3–Hydroxy–3–Methylglutaryl–CoenzymeA Reductase. J Clin Invest 76:1748–1754

Maltese WA, Sheridan KM, Repko EM, Erdman RA (1990) Post–translational modification of low molecular mass GTP–binding proteins by isoprenoid. J Biol Chem 265:2148–2155

Maltese WA, Sheridan KM (1985) Differentiation of neuroblastoma cells induced by an inhibitor of mevalonate synthesis:relation of neurite outgrowth and acetylcolinesterase activity to changes in cell proliferation and blocked isoprenoid synthesis. J Cell Physiol 125:540–558

Paoletti P, Butti G, Knerich R, Gaetani P, Assietti R (1990) Chemothereapy for malignant gliomas of the brain: a review of ten–years experience. Acta Neurochir 103:38–46

Soma MR, Pagliarini P, Butti G, Paoletti R, Paoletti P, Fumagalli R (1992) Simvastatin, an inhibitor of cholesterol biosynthesis, shows synergistic effect with N,N'–Bis(2–chloroethyl)–N–nitrosourea and beta interferon on human glioma cells. Cancer Res 52:1–9

CNS AND CARDIOVASCULAR EFFECTS OF TNF-α

F. Squadrito, A.P. Caputi
Institute of Pharmacology
School of Medicine
University of Messina
Piazza XX Settembre 4, 98121
Messina ITALY.

Recombinant cytokines have high potential for modulating immunity in cancer, viral illness and immunodeficient states. However their large use in clinical trials turned out to be impraticable because of severe systemic side-effects (Chouaib et al., 1992). The cardiovascular system and Central Nervous System (CNS) represent the main targets of Tumor Necrosis Factor (TNF-α) toxicity. Experimental studies have confirmed that systemic or intracerebral injection of recombinant TNF-α has profound cardiovascular and neurological effects in addition to its pyrogenic and metabolic activity. Understanding and elucidating the mechanisms underlying cardiovascular and CNS toxicity should enable the design of molecules to oppose this toxicity without interfering with the immune reactions.

CNS EFFECTS OF TNF-α

Patients treated with recombinant cytokines such as recombinant human TNF-α (rhTNF-α) are feverish, and complain of anorexia, fatigue and general malaise. Higher doses of recombinant cytokines lead to emotional and psychiatric symptoms ranging from depression to irritability and delirium (Kent et al., 1992). Neurotoxicity can be induced by both peripheral and central administration of recombinant cytokines. However these molecules are more potent when they are injected centrally than via the systemic administration. These data indicate that these effects are mediated directly in the CNS. Such an interpretation requires an understanding of the ways by which peripherally injected cyokines can enter the brain. Cytokines are large, hydrophilic proteins. Consequently, they would not be expected to cross the blood-brain barrier in significant amounts without the aid of a transport system.

Although such a possibility has been proposed (Kent et al., 1992), it has been speculated that cytokines do not enter the brain, but act at the level of the circumventricular organs where the blood barrier is non-existent. The target circumventricular organ for cytokines would be the organum vasculosum laminae terminalis (OVLT), located at the anterior wall of the third ventricle. The concept is that TNF-α binds to cell located on the vascular side of OVLT, thereby inducing a release of second mediators (i.e. prostaglandins, platelet activating factor, nitric oxide), which then freely diffuse to nearby hypothalamic regions.

In experimental animals locomotor activity and expecially food intake were used as indexes

NATO ASI Series, Vol. H 75
Cancer Therapy
Edited by N. D'Alessandro, E. Mihich, L. Rausa,
H. Tapiero, and T. R. Tritton

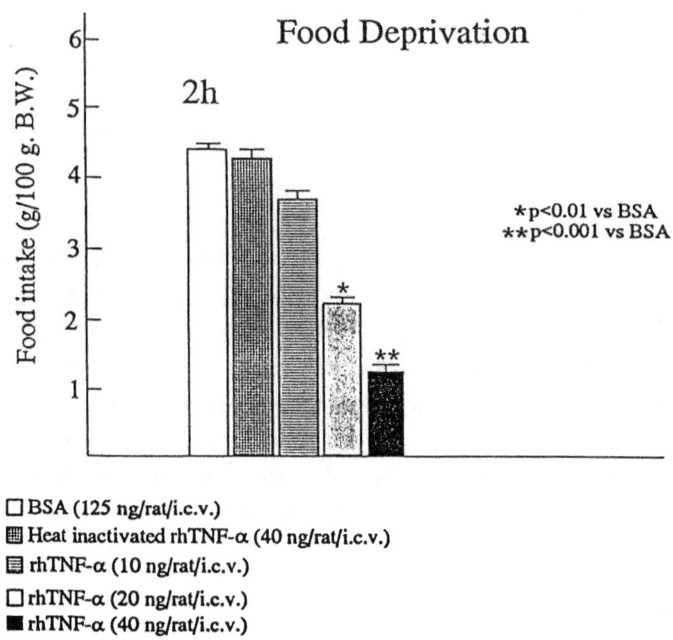

Figure 1 - Effects of rh-TNF-α on food deprivation.

of neurotoxicity. However food intake represent only one aspect of a more complex behavior called "ingestive bahvior". Ingestive behavior include, in fact, feeding behavior and drinking behavior. We therefore will discuss separately these two features of ingestive behavior which were used in our experimental work as "tools" to investigate the mechanisms underlying TNF-α neurotoxicity.

Feeding behavior

A large body of evidence indicates that TNF-α may be involved in several CNS functions: in fact TNF-α has been detected in the brain and has been reported to cause astroglial surface expressions of major histocompatibility complex (Vilcek and Lee, 1991). To investigate the effects of TNF-α on feeding behavior, we used male Sprague-Dawley rats that were deprived of food but not of water for 24 h. Intracebroventricular administration of recombinant human TNF-α (rhTNF-α ; 10, 20 and 40 ng/icv) produced a dose-dependent reduction in food intake (Figure 1). When the highest dose of rhTNF-α (40 ng) was injected intraperitoneally, it failed to exert any anorectic effect.

Platelet activating factor (PAF) is an endogenous phosholipid which has been implicated as a mediator of allergic, inflammatory and shock processes (Squadrito et al., 1991).

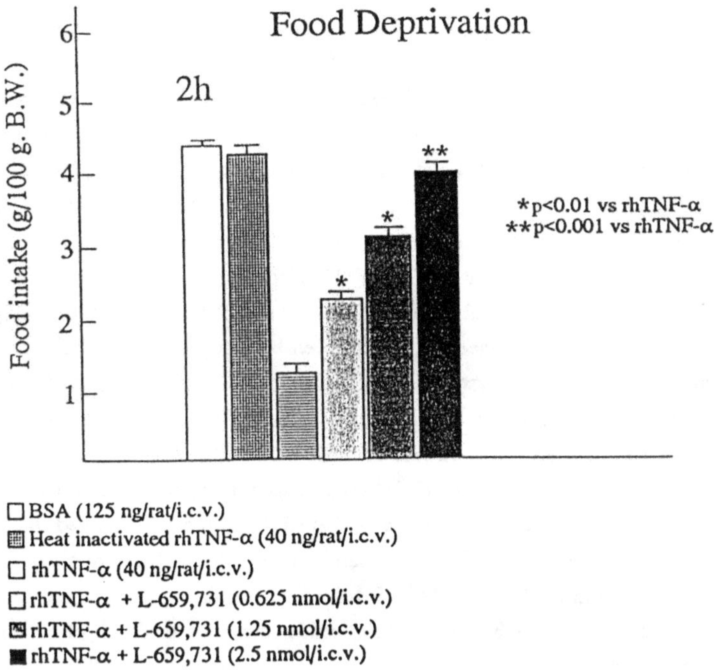

Figure 2 - Effects of L-659,731 on rh-TNF-α induced anorexia.

It has also been demonstrated that PAF is produced by various cell types including brain cells (Plata-Salaman, 1988); in addition high affinity binding sites for PAF have been shown in the rat brain which modulate specific CNS functions. Experimental findings have suggested that intracerebroventricular microinfusion of PAF suppresses food intake in rats (Plata-Salaman, 1988). Furthermore it has been shown that PAF, in vitro, stimulates the release of TNF-α by uman and rodent lymphocytes and macrophages (Rola-Pleszczyski et al., 1988). In light of the above findings we carried out experiments to investigate whether TNF-α induced anorexia could be mediated, at least in part, by PAF. Figure 2 shows that intracerebroventricular administration of L-659, 731 , a specific PAF receptor antagonist, significantly abolished rhTNF-α induced inhibition of food deprivation. These data collectively indicate that TNF-α neurotoxicity could be mediated by PAF and that PAF receptor antagonists might be effective in opposing this neurotoxicity.

Drinking behavior

Drinking behavior is a very complex process (Fueller, 1984). Thirst may be affected by a variety of mechanisms, including the direct action of "thirst neurones", release of substances which affect these neurones or altered water or electrolyte balance in the body, causing a secondary stimulation or inhibition of the intake of water (Grossman, 1990). Water deprivation represents a physiological stimulus that leads to cellular and extracellular dehydration and consequently to ingestion of an amount of water capable of balancing fluid and electrolytes. Spontaneous or aquired behavior such as drinking behavior induced by water deprivation may be used as a tool to investigate the neurotoxicity of several molecules since this phenomenon is centrally mediated (Calapai et al., 1990). In rats deprived of water, rhTNF-α (40 ng/rat) causes a profound and significant inhibition of water intake (Squadrito et al., 1992). TNF-α may induce the release of nitric oxide (Ding et al. 1988). Furthermore nitric oxide synthase has been also shown to be present in the CNS (Knowles et al., 1990) and astrocytes release nitric oxide in response to several stimuli (Murphy et al., 1990). As far as the biological function of nitric oxide in the brain is concerned, it has been been suggested that nitric oxide acts as an inhibitor mechanism when thirst is stimulated by water deprivation (Calapai et al., 1992). These data, taken togheter, led to the hypothesis that rhTNF-α induced inhibition of drinking behavior could be mediated by the release of nitric oxide. This hypothesis has been confirmed using n-nitro-L-arginine methyl ester (L-NAME), a specific inhibitor of nitric oxide synthesis. In fact intracebreventricular administration of L-NAME significantly restored the impairment in drinking behavior induced by rhTNF-α (Squadrito et al., 1992). These findings suggest that the nitric oxide pathway in the brain might be involved in mediating the neurotoxicity of TNF-α and indicate that specific inhibitors of nitric oxide synthesis could be of benefit in order to blunt TNF-α neurotoxicity.

CARDIOVASCULAR EFFECTS OF TNF-α

Systemic administration of rhTNF-α results in hypothesion, tachycardia, and tachypnea (Kettelhut et al., 1987)). Moreover rhTNF-α administration may induce myocardial dysfunction. Inherent in the pathophysiology of septic shock is a severe decrease in myocardial performance associated with biventricular dilatation, depressed ejection fractions, alteration in Frank-Starling relationships and profound serial decreases in cardiac index, which ultimately may be fatal (Cunnion and Parrilo, 1989). This myocardial dysfunction results from circulating myocardial depressant factors (MDF), which are present in the sera of patients with sepsis (Comstock and Thomas, 1989). TNF-α may itself be a circulating MDF; infusion of rhTNF-α into dogs leads to a depressed cardiovascular profile (Kettelhut et al., 1987).
Also rhTNF-α blocks the intramyocyte accumulation of cAMP and decrease the myocyte contractile response to β-adrenergic stimulation (Gulick et al., 1989). Finally elevated circulating levels of TNF-α have been noted in chronic heart failure (Levine et al., 1990). All these findings, taken togheter, clearly indicate that TNF- α may impair myocardial function through different mechanisms.

Leukocyte adherence and accumulation may also cause myocardial injury. Following inflammation and ischaemic states leukocytes are localized along the vessel wall where they adhere to endothelial cell, cross the endothelial barrier through interendothelial cell junctions and accumulate in the subcellular regions (Osborn, 1990). Leukocytes may, in turn, produce irreversible changes (cell death of myocytes) by releasing several damaging and harmful substances (Butcher, 1990). Ahesion molecules are necessary for leukocytes accumulation in the myocardial. Endothelial leukocyte adhesion molecules-1 (ELAM-1) are needed for the initial binding and accumulation of neutrophils, while intercellular adhesion-molecules-1 (ICAM-1) are important for the migration and extravasation of leukocytes. The expression of these molecules may be induced in vitro by the addition of rhTNF-α (Mantovani and Dejana, 1989). In addition local injection of rhTNF-α causes a rapid recruitment of leukocytes from the blood and this cytokine can be detected at sites of inflammatory lesions. Therefore these findings indicate that TNF-α can induce cardiotoxicity by enhancing the myocardial accumulation of leukocytes. In agreement with this hypothesis, it has been recently suggested that elevated circulating levels of TNF-α are present during myocardial ischaemia-reperfusion injury and a passive mmunization with specific antibodies rised against TNF-α reduces myocardial damage and polymorphonuclear leucocyte accumulation following coronary artery occlusion and reperfusion in the rat (Caputi and Squadrito, 1992). These data therefore explain the mechanisms underlying TNF-α induced cardiotoxicity and also led to the hypothesis that the cytokine is an important mediator of myocardial reperfusion injury.

CONCLUSIONS

Besides inducing positive effects on immunity, the administration of rhTNF-α is associated with a severe cardiovascular and neurological toxicity. In the CNS TNF-α induced toxicity seems to be mediated by the release of second mediators such as platelet activating factor and nitric oxide. TNF-α also impairs cardiovascular system by inducing both a severe hypothension and a myocardial dysfunction by several mechanisms. Thes findings suggest possible means for reducing the toxicity of administered rhTNF-α and for improving therapy in immunodeficient states.

REFERENCES

Butcher EC (1990) Cellular and molecular mechanisms that direct leukocyte traffic. Am J Pathol 136: 3-11.

Calapai G, Squadrito F, Massi M, Caputi AP, de Caro G (1990) Endotoxin inhibition of drinking behavior in the rat. Pharmacol Res 22: 161-170

Calapai G, Squadrito F, Altavilla D, Zingarelli B, Campo GM, Cilia M, Caputi AP (1992) Evidence that nitric oxide modulates drinking behavior. Neurophamacol 31: 761-764.

Caputi AP, Squadrito F (1992) Role of TNF- α and therapeutic perspectives in the model of bowel and myocardial ischaemia/reperfusion injury. Pharmacol Ress 26 (suppl 2): 150-151.

Comstock LE, Thomas DD (1989) Penetration of endothelial cells monolayers by Borrelia burgdorferi. Infect Immun 57:1626-1628.

Chouaib S, Branellec D, Buurman WA (1991) More insights into the complex physiology of TNF. Immunol Today 12: 141-142.

Cunnion RE, Parrillo JE (1989) Myocardial dysfunction in sepsis. Crit Care Clin 5: 99-118.

Fuller LM (1984) The pharmacology of drinking behavior. Pharmacol Ther 24: 179-206

Grossman SP (ed) (1990) Thirst and Sodium appetite. Physiological basis. Academic Press San Diego

Gulick T, Chung MK, Pieper SJ, Lange LG, Screiner GF (1989) Interleukin 1 and tumor necrosis factor inhibit cardiac myocyte β-adrenergic responsiveness. Proc Natl Acad sci USA 86: 6753-6757.

Kent S, Bluthe' R, Kelley KW, Dantzer R (1992) Sickness behavior as a new target for drug development. TiPS 13: 23-28.

Kettelhut IC, Fiers W, Goldeberg L (1987) The toxic effect of tumor necrosis factor in vivo and their prevention by cyclooxygenase inhibitors. Proc Natl Acad Sci USA 84: 4273-4277.

Knoweles RG, Palacios M, Plamer RM, Moncada S (1990) Kinetic characteristics of nitric oxide from rat brain. Biochem J 269: 207-210

Levine B, Kalman J, Mayer L, Fillit HM, Packer M (1990) Elevated circulating levels of tumor necrosis factor in in severe chronic heart failure. N Engl J Med 323: 236-240.

Mantovani A, Dejana E (1989) Cytokines as communication signal between leukocytes and endothelial cells. Immunol Today 10: 370-375.

Murphy S, Minor RL, Welk G, Harrison DG (1990) Evidence for an astrocyte-derived vasorelaxing factor with properties similar to nitric oxide. J Neurochem 55: 349-351.

Osborn L (1990) Leukocyte adhesion to endothelium in inflammation. Cell 62: 48-51.

Plata-Salaman CR (1988) Food intake suppression by immunomodulators. Neurosci Res Comm 3: 159-165.

Rola-Pleszczyshi M, Bosse J, Bissonnette E, Dubois C (1988) PAF acether enhances the production of tumor necrosis factor by human and rodent lymphocytes and macrophages. Prostaglandins 35: 802-806.

Squadrito F, Sturniolo R, Altavilla D, Santoro G, Campo GM, Arena A, Caputi AP (1991) Platelet activating factor involvement in splanchnic artery occlusion shock. Eur J Pharmacol 192: 47-50.

Squadrito F, Calapai G, Zingarelli B, Altavilla D, Campo GM, Caputi AP (1992) Evidence for a role of nitric oxide in the endotoxin inhibition of drinking behavior in the rat. In : Moncada S, Marletta MA, Hibbs JB, Higgs RA (eds) The biology of nitric oxide I Portland Press London, 276.

Vilcek J, Lee TH (1991) Tumor necrosis factor: new insights into the molecular mechanisms of its multiple actions. J Biol Chem 266: 7313-7316.

ANGIOGENESIS AND ANGIOGENESIS FACTORS IN STAGES OF CARCINOGENESIS

Claudio J. Conti
Department of Carcinogenesis
The University of Texas
M.D. Anderson Cancer Center
Science Park-Research Division
P.O. Box 389
Smithville, Texas 78957

Over the last decade, we have witnessed a dramatic advance in the knowledge of the intimate mechanisms leading to cancer development. New methodologies have allowed the identification and cloning of genes involved in carcinogenesis as well as the determination of cell signalling pathways, cell-to-cell interactions, and agonistic and antagonistic factors determining cellular behavior (i.e. proliferation, differentiation, migration, secretion etc.). In the near future, the knowledge accumulated over the last decade will likely be instrumental in the design of new preventive and therapeutic approaches to cancer.

In this chapter, I review the concept of multistage carcinogenesis and describe the importance of angiogenesis as one of the critical stages in tumor development. I also stress the importance of angiogenesis as a target for new therapeutic approaches.

THE MULTISTAGE NATURE OF CARCINOGENESIS

The multistage nature of cancer development is now widely accepted. This concept was supported by two early lines of evidence: the observation by pathologists that the vast majority of cancer develops from premalignant lesions and investigations in animal models using multistage carcinogenesis protocols.

For decades pathologists have described lesions with high probability of malignant transformation and have considered these lesions as precursors of cancer (Reviewed in Sirica, 1989 and Henson and Albores-Saavedra, 1993). A good example of such lesions are cervical dysplasias and carcinomas in situ. Other well-known examples are solar keratosis and Bowen's disease, ductal dysplasias of breast and

NATO ASI Series, Vol. H 75
Cancer Therapy
Edited by N. D'Alessandro, E. Mihich, L. Rausa,
H. Tapiero, and T. R. Tritton

prostate, and some colon adenomas. (Henson and Albores-Saavedra, 1993).

The other line of evidence for multistage carcinogenesis derives from animal experiments. Early work performed in mouse skin showed that skin tumors (papillomas and squamous carcinomas) can be induced by a single subcarcinogenic dose of a carcinogen followed by multiple application of a non-carcinogenic but strongly irritating agent. (Reviewed in Slaga, 1989 and DiGiovanni, 1992, Conti and Klein-Szanto 1992). These experiments suggested that carcinogenesis consisted in at least two stages, which were termed initiation and promotion. It was postulated that initiation, the first stage, occurs as the direct geneotoxic action of the carcinogen on the epidermal cells. This stage produces nonvisible phenotypic change but is irreversible. In contrast, the second stage, promotion, requires that the "promoting" agent be applied continually over time, and its effects are reversible. The concept of initiation-promotion has been extended to other animal systems and (e.g., in liver, bladder, and colon), and epidemiological evidence suggests that similar mechanisms are responsible for cancer in humans. (Peraino and Jones, 1989).

CELLULAR AND MOLECULAR BASES FOR MULTISTAGE CARCINOGENESIS

Beremblum's early work in the mouse skin system showed that the initiation event was irreversible and accumulative and perhaps caused by genetic damage. In contrast, promotion was shown to be reversible and likely to be produced by epigenetic mechanisms (Peraino and Jones, 1989; DiGiovanni, 1992).

Work done by several laboratories in the early 1980s provided the basis for interpreting the nature of the initiation event. Investigations carried out in two chemical carcinogenesis animal models showed that the initiation event was probably related to the mutation of a gene in the ras family (Quintanilla et al., 1986; Sukumar et al., 1983). These mutations appeared to provide a growth advantage to the initiated cells; it was thought, therefore, that initiated cells would expand clonally if stimulated by either endogenous factors (estrogen in the rat mammary gland) or exogenous factors (promoter in the mouse skin), thus generating a premalignant lesion (Yuspa and Poirer, 1988). We know now that the initiation-promotion model is probably an oversimplification, and current evidence suggests that multiple genetic events are required for fully malignant conversion (Balmain and Brown, 1988), but the overall idea of the model is still fully valid: cells acquire a mutation that confers a growth advantage and consequently they expand, becoming targets for further mutation (Figure 1). However, the present view of carcinogenesis is that this processed is repeated several times during the natural history of tumor development, and as a result of this multistage process. more and more aggressive clones develop until cancer arises. (Nowell, 1986; Peraino and Jones, 1989).

These models of cancer development are by no means limited to animal experimental systems. Epidemiological studies of human cancer are consistent with these models and, furthermore, investigations in the last few years in several human tumors have clearly shown the presence of multiple genetic events in the cancer cell (Vogelstein et al., 1988; Weinberg, 1989; Bishop, 1991). Research into the mechanisms of carcinogenesis over the last decade has not only shown the importance of the accumulation of genetic damage but also made possible the identification and cloning of several genes that appear to be responsible for these genetic alterations. Cancer genes can be divided into two major families: oncogenes and tumor suppressor genes.

Oncogenes are cellular genes that play a role in regulating normal cellular function. However, the overexpression, mutation, or dysregulation of these genes can contribute to carcinogenesis. They are considered to be dominant because the activation of a single allele is sufficient to confer oncogenic properties. The product of oncogenes have been shown to have different cellular functions: growth factors (e.g. sis), growth factor receptors (e.g. erb, neu), signal transduction (e.g. ras), regulator of transcription (e.g. fos, jun) (Weinberg, 1989; Bishop, 1991).

In contrast to oncogenes, tumor suppressor genes are recessive, and thus are considered negative regulators of carcinogenesis: both alleles must be inactivated to tumor development. Only a few suppressor genes have been cloned and fully characterized, two of which are the retinoblastoma gene, which appears to be involved in the regulation of the cell cycle, and the p53 gene, which appears to be a transcription factor. Other putative suppressor genes have been recently cloned and are being characterized (Knudson, 1986; Sager, 1986; Stanbridge, 1990).

In summary, the most accepted model of carcinogenesis is that malignancy is the result of accumulation of genetic damage, i.e. the activation of oncogenes and the inactivation of tumor suppressor genes. In colon cancer, for example, it has been postulated that carcinogenesis may be due to the activation of one oncogene (e.g. K-ras) and the inactivation of at least three tumor suppressor genes (MCC, p53, and DCC). In lung cancer cells, activation of one oncogene, K-ras, and the inactivation of multiple tumor suppressor genes (p53, retinoblastoma gene, and a putative gene mapped in chromosome 3p are common findings.

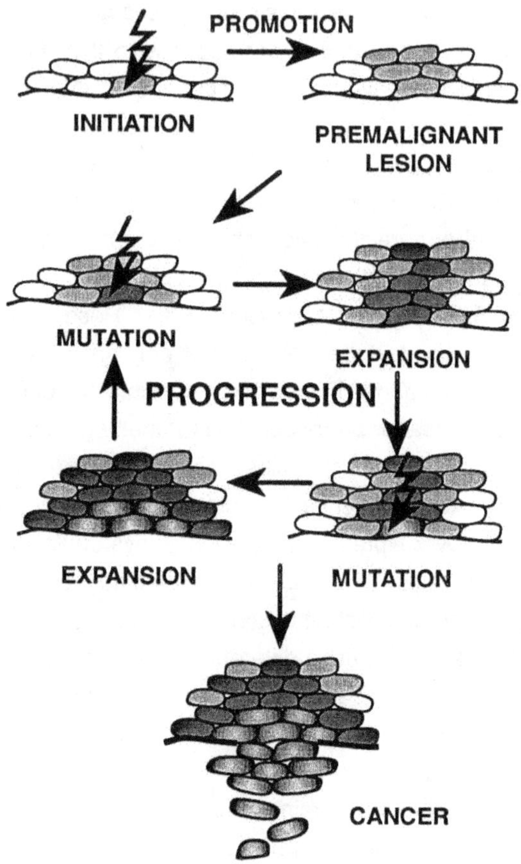

Figure 1: Schematic representation of the cellular and genetic events leading to cancer development.

WHY ARE MULTIPLE STAGES IN CANCER DEVELOPMENT NECESSARY?

The hyposthesis that multiple genetic events are required for carcinogenesis is based on two precepts:

1. Cells have several redundant regulatory mechanisms that need to be altered to allowed unrestricted proliferation. Therefore, multiple genetic events are required either to dysregulate or to abrogate those mechanisms.
2. Cancer is not only the result of cell proliferation, but the result of multiple abnormal behaviors. The most important of them are:

a. Unrestricted proliferation
b. Loss of balance between proliferation vs differentiation or apoptosis
c. Autonomous growth
d. Induction of angiogenesis
e. Invasion to adjacent tissues
f. Access to the blood stream and colonization of other organs (metastasis)

As can be seen, only points a,b, and c pertain to cell proliferation whereas points d,e and f are more related to the capacity of the tumor cell to modify the environment. All of these behaviors are likely regulated by a number of different genetic mechanisms. Therefore, it is reasonable to expect that multiple genetic alterations will be required to develop a fully malignant phenotype.

In the next sections, we will discuss in more detail the importance of vascularization to cancer development, as well as the possible mechanisms involved in tumor angiogenesis.

TUMOR ANGIOGENESIS

It is widely accepted that tumor development depends on the formation of new blood vessel (angiogenesis). Because premalignant or malignant lesions cannot grow beyond a certain size (approximately 1 mm^3) without angiogenesis (Folkman, 1990; Liotta et al, 1974). Angiogenesis appears to be a requirement for the transition between in situ to fully invasive lesions. Recent experiments in transgenic mice have supported this concept. Folkman et al. (1989), investigate a transgenic mice in which an oncogene was put under the regulatory region of the insulin gene. The resultant mice develop hyperplasia of the pancreatic ß cells. However, only few of these hyperplastic lesions progress to carcinomas. Interestingly, angiogenesis precedes the onset of neoplastic transformation, indicating that angiogenesis is a key element in the transition from the hyperplastic lesion to neoplasia.

Other animal models have also shown the importance of vascularization during early stages of carcinogenesis. Polverini and Solt (1988), cultured a subpopulation of keratinocytes with angiogenic properties from carcinogen-treated cheek hamster epithelium. This subpopulation of keratinocytes was able to induce neovascularization in rat corneas and migration of endothelial cells in cultures from bovine adrenal glands. Interestingly, the angiogenic keratinocytes can be isolated from the epithelium several weeks before the onset of malignant lesions. The same researchers, using cell fusion experiments, have confirmed that angiogenesis is an early phenotypic trait and have postulated that it is under the control of a tumor suppressor gene (Polverini et al, 1988; Moroco et al, 1990).

Angiogenesis has also been shown to play a role in metastasis. A correlation between the extent of vascularization and the metastatic properties of a tumor have been shown in breast carcinoma (Weidner et al, 1991; Bosari et al, 1992) and in non-small cell lung cancer (Macchiarini et al. 1992). Two mechanisms may explain this correlation. Newly formed capillaries in the tumor may have a fragmented basement membrane, therefore facilitating the access of tumor cells to the blood stream, and induction of new blood vessels may be essential to the development of a tumor at the site of metastasis.

HISTOGENESIS OF BLOOD VESSELS

The sprouting model (Ausprunk, 1979) which is based on studies of experimental wounds, describes angiogenesis (Figure 2). Angiogenesis seems to originate from existing blood vessels, mainly small veins. Probably the first event is the secretion of proteolytic enzymes (e.g. collagenases, plasminogen activator) that degrade the basement membrane. Endothelial cells then migrate to the perivascular tissues and proliferate, forming a "dead end" hollow cord, i.e. one that is closed on one end but open on the other. Eventually the newly formed capillaries anastomose, forming loops that carry blood. The processes involved in the reconnection of blood vessels is poorly understood (Ausprunk and Folkman, 1977; Rev. in Montesano, 1992 and Arnold and West, 1991).

Regulation of angiogenesis is probably very complicated and appears to involve several mechanisms. Furthermore, the whole process appears to respond to a very delicate balance between proteases and antiproteases, as well as between proliferation and migration. In this regard, it is important to note that endothelial cells are prone to contact inhibition, and, therefore, proliferation does not occur in a normal continuous endothelium. Unlike cells from other epithelia, endothelial cells cannot proliferate and migrate to a distant location by displacing themselves over other endothelial cells. Therefore, most of the activity in terms of proliferation, migration, and remodeling, which result in capilar tubes, occurs exclusively in the dead-end of the new capillary tubes. (Montesano, 1992; Arnold and West, 1991).

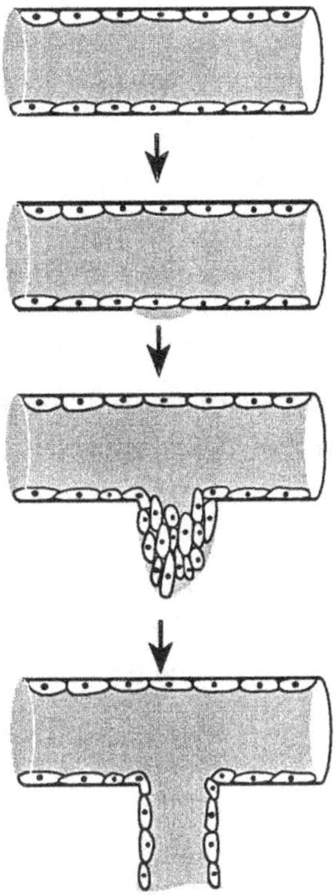

Figure 2: Histogenesis of blood vessels.

INDUCTION OF ANGIOGENESIS BY TUMORAL CELLS

The induction of vascularization by tumors is still not completely understood and appears to involve several concurrent mechanisms. A substantial body of evidence has clearly shown that tumor cells can produce soluble factors that directly or indirectly affect tumor angiogenesis (Rev. in Folkman and Klagsbrun, 1987, D'Amore and Klagsbrun, 1989 and Rissau, 1990). However, the complexity of angiogenesis regulation, as well as the possibility that multiple alternative paths of angiogenesis may be operational in different tumors, has delayed the possibility of building a consensus theory of tumor angiogenesis. It is also very important to consider that the angiogenic activity may not necessarily be a direct effect on the endothelial cell. Current evidence suggests that some tumor angiogenic factors may be mediated by stromal cells. Also, it is possible that angiogenic factors secreted by tumor cells may modify the extracellular matrix, creating favorable conditions for the sprouting of new capillaries. (Arnold and West, 1991; Rissau, 1990; Klagsbrun and D'Amore, 1991).

Early experiments in the 1960s showed that angiogenesis factors extracted from tumors can diffuse through 0.45 µm filters, (Greenblatt and Shubik, 1968; Ehrman and Knoth, 1968). Since then, several putative angiogenic factors have been identified, and some of them have been characterized. (Rev. in Rissau, 1990, Folkman and Klagsbrun, 1987, and Blood and Zelter, 1990).

Angiogenic factors are not necessarily mitogenic for endothelial cells. As discussed in a previous section, at least three basic functions are required in neovascularization, i.e. proteolysis (to degrade basement membrane and also to remodel sprouts), migration, and proliferation. Therefore, the angiogenic factors can modulate any of these functions. For example, several angiogenic factors have been shown to increase endothelial cell motility without inducing proliferation. Others are believed to act by recruiting inflammatory cells in the tumor stroma.

SOLUBLE ANGIOGENIC FACTORS

In the present section we will briefly describe some of the more likely candidates for tumor-related angiogenesis. A more detailed description of these factors can be found in recent extensive reviews (Folkman and Klagsbrun, 1987; D'Amore and Kagsbrun, 1990; Rissau, 1990; Klagsbrun and D'Amore, 1991; Blood and Zelter, 1990).

Fibroblast Growth Factor: FGF is a family of growth factors, each of which having a molecular weight of 18000. The first members of this family to be isolated were acidic FGF (aFBF) and basic FGF (bFGF). These forms are 53% homologous, bind heparin, and are growth factors for a wide variety of cells, including endothelial cells in vivo and in vitro. aFGF is normally found in brain and retina, whereas bFGF is widely distributed. Interestingly, because these growth factors lack a signal peptide, they are not secreted. It has been postulated that this growth factor might be released when cells are damaged, generating tissue repair. Alternatively, it has also been suggested that FGF may be stored bound to the basement membrane and is released by tumor enzymes. More recently, molecules with high homology with FGF has been described. These molecules contain signal peptides and therefore can be secreted. Two of these molecules are the products of two oncogenes: int-2 and hst-1 which are the site of integration of murine mammary tumor virus.

Transforming Growth Factor α (TGFα) and Epidermal Growth Factor (EGF): TGFα and EGF bind to the same receptor, EGFr. Primarily isolated from virally transformed rodent cells, TGFα is overexpressed in a variety of tumors, in which it is considered an autocrine or paracrine factor. EGF is a well-studied cytokine that bears 35% homology with TGFα and binds the same receptor. However it is approximately tenfold less active in angiogenesis assays. Both TGFα and EFG stimulate angiogenesis in vivo and are mitogenic for endothelial cells in vitro.

Transforming Growth Factor ß (TGFß): TGFß is a family of peptides originally isolated from transformed rodent cells. Interestingly, although TGFß is a growth inhibitory factor for most epithelial cells, it is overexpressed in several tumors. This paradoxic behavior may be explained by the fact that TGFß induces angiogenesis in vivo. However, TGFß does not promote endothelial proliferation or migration in vitro, but it influences the formation of the endothelial lumen once the endothelial cells have divided and migrated. It is also possible that TGFß, may have an angiogenic action through other cell types. This growth factor has a potent chemotactic action for macrophages.

Angiogenin: Angiogening is a polypeptide with a molecular weight of 14000, was originally isolated from a human adenocarcinoma cell line. Although it has some homology with a family of ribonucleases, it is unlikely that the ribonuclease activity is related to its angiogenic activity. Interestingly, this is another compound that is angiogenic in vivo. but apparently has no direct action on endothelial cells. Therefore, it is likely that it may have indirect effect on vascularization.

Angiotropin: Angiotropin has a molecular weight of 4500 and is isolated from peripheral blood monocytes. It is a potent angiogenic factor in vivo and induces endothelial cell migration in vitro. However, it does not appear to be mitogenic for endothelial cells.

Vascular Endothelial Growth Factor (VEGF): Although VEGF binds heparin, it is unrelated to other members of the FGF family. It has a molecular weight of 45,000 and is isolated from bovine pituitary cells and tumor cell lines. It produces angiogenesis in vivo and is a specific mitogen for endothelial cells in vitro. It also increases vascular permeability.

Prostaglandins: Prostaglandins are well-known cell mediators and some of them also have angiogenic activity. These mediators of inflammation and other tissue functions are derived from arachidonic acid stored in cell membranes. They are produced by a wide variety of cell types.

Platelet derived Endothelial Cell Growth Factor (PD-EGF): PD-EGF has a molecular weight of 45,000. It is responsible for most of the vascularization properties of platelets and is produced by epithelial cells and carcinomas. Although unrelated to the FGF family, PD-EGF, like the FGFs, lacks signal peptide. Therefore PF-EGF is not secreted but probably requires cell lysis to induce angiogenesis. Its mechanisms of action are not clear, but it produces endothelial cell proliferation in vivo and in vitro.

Tumor Necrosis Factor α (TNFα): TNFα is produced by macrophages and some tumor cells and has a molecular weight of 17,000. This polypeptide has a paradoxic effect similar to that of TGFβ. Although it inhibits endothelial cells, it is angiogenic in in vivo assays.

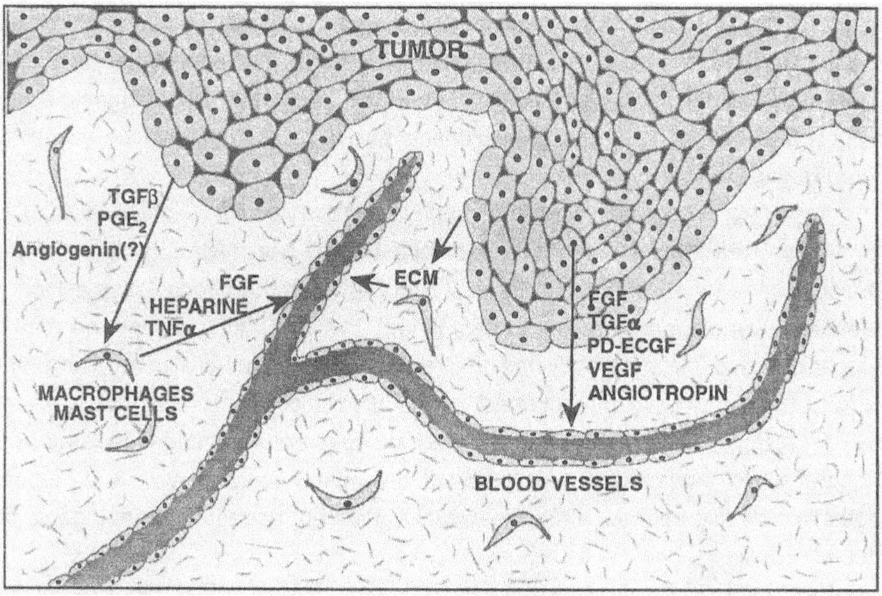

Figure 3: Interaction between tumor cells and blood vessels.

RECESSIVE-NEGATIVE MECHANISM OF ANGIOGENESIS CONTROL

In the previous section, several positive factors of angiogenesis were described. This model assumes that the endothelium is normally non- proliferative and vascularization does not occur unless stimulated by specific factors. Although this model is generally accepted, some authors have suggested that a negative recessive factor may be a key control in angiogenesis (Moroco et al., 1990). This alternative model postulates that angiogenesis is constitutive in tissues but normally controlled by recessive suppressor factors analogous to the protein products of tumor suppressor genes i.e. both copies of the gene need to be inactivated to produce angiogenesis (Sager, 1986). The evidence of a model of a negative model of angiogenesis is derived from the hamster cheek pouch studies carried out by Polverini and Solt (1988). Using this animal model, these authors identified an anti-angiogenesis factor in conditioned media of tumor-suppressed hamster-human hybrids. (Moroco, et al., 1990). The inhibitor appeared to be homologus to thrombospondin, a glycoprotein found in platelets (Rastinejad et al., 1989; Good et al., 1990).

In addition to thrombospondin several other inhibitors have been described, including platelet factor IV, protamine and several inhibitors isolated from non-vascularized tissue (Klagsbrun and D'Amore, 1991).

MULTIPLE FACTORS AND MECHANISMS OF ANGIOGENESIS

The identification of several positive and negative regulators of angiogenesis raised the question of the purpose of this apparent redundancy. It is possible that not all these factors are actually involved in angiogenesis in vivo. The artificial nature of the experimental system may have detected factors that do not have an angiogenic activity in real situations. Also it is possible that different tumors use different pathways to induce angiogenesis. The same would be true for wound repair, in which different angiogenesis pathways ensure that the repair mechanisms are initiated in any circumstances and by a variety of tissues capable of detecting the need for repair.

The most likely explanation at this point is that angiogenesis takes place when the normal balance of positive and negative regulators is broken as result of trauma, inflammation, or the presence of a tumoral cell secreting a variety of cytokines and growth factors.

ANGIOGENESIS AS A TARGET OF CHEMOTHERAPY

For several reasons, angiogenesis has been considered a target for chemotherapeutic approaches:

1. It has been shown that angiogenesis is essential for invasion and metastasis.
2. Angiogenesis is mediated by extracellular factors, which are relatively easy targets of therapeutic agents.
3. Angiogenesis is normally suppressed in adult tissue, and therefore inhibition of angiogenesis is unlikely to produce side effects.

EXPERIMENTAL INHIBITION OF ANGIOGENESIS

Several laboratories have shown that it is possible to inhibit angiogenesis in experimental models.

One early approach was to examine avascular tissues for angiogenesis inhibitors. With this approach, angiogenesis was inhibited either by cartilage pieces placed in the vascular bed (Eisenstein et al., 1973, Brem and Folkman, 1975) or by purified factors obtained from cartilage (Langer, et al., 1980). The cartilage angiogenesis inhibitors appear to have collagenase inhibitor properties (Murray et al., 1986). Other avascular tissues such as the eye lens and vitreous have also been shown to produce angiogenesis inhibitor factors (Williams, et al., 1984; Taylor and Weiss, 1985).

Probably the most studied approach of angiogenesis inhibition has been the use of steroids. Cortisone, either alone or in combination with heparin was shown to inhibit tumor angiogenesis (Taylor and Folkman, 1982; Heuser et al., 1984). However, different laboratories have reported different results in tumor regression after angiogenesis inhibition. While Folkman et al. (1983) reported tumor regression, other authors like Ziche et al., (1985) and Sakamoto et al., (1987), have observed tumor growth inhibition but not regression.

Several other inhibitors of angiogenesis have been reported in the last several years. They include GPA 1734, an inhibitor of basement membrane synthesis (Missirlis et al., 1990), suramin, a trypenocydal agent shown to block the binding of several growth factors (Gagliardi, et al., 1992) depletion of copper (Brem et al., 1990) inhibitors of protaglandin synthesis, and heparin antagonists among others (reviewed in Blood & Zetter, 1990).

The inhibition of angiogenesis in experimental models has shown the feasibility of an antineoplasic therapeutic approach based on inhibition of angiogenesis. However, these results also suggest that inhibition of angiogenesis will stop tumor growth rather than produce tumor regression. Therefore, it is likely that angiogenesis inhibition will be useful either in combination with other therapies or as postsurgical therapy for the prevention of recurrence and metastasis.

SUMMARY

It has been clearly established in carcinogenesis animal models, as well as in human tumors, that cancer development is a multistage process. These stages of carcinogenesis most likely represent different levels of accumulation of genetic damage, mainly activation of oncogenes and inactivation of tumor suppressor genes.

The need for multiple genetic events may be due to two main reasons: the existence of redundant mechanisms that regulate cellular functions and the need for multiple phenotypic changes during the acquisition of neoplastic behavior. Among these phenotypic changes are the capacity of unrestricted proliferation, autonomous growth, induction of angiogenesis, and the potential to invade other tissues and metastasize.

The capacity of inducing vascularization (angiogenesis) has been recognized for many years as one of the essential characteristics of neoplastic cells. It is well established that without vascularization tumors cannot grow larger that 1-2 mm^3. Experimental models have shown that vascularization appears to occur at early stages of carcinogenesis, before the preneoplastic hyperplastic lesion becomes fully malignant. Recent investigations have also shown a correlation between vascularization and metastatic potential of a tumor.

Some aspects of the mechanisms of angiogenesis are presently being investigated. Previous investigation have shown that tumors secrete diffusible factors that may activate angiogenesis. Several potential angiogenesis factors have been isolated and characterized. Among them the best studied are: angiogenin, FGFs, TGFα, TGFβ, TNFα, angiotropin, VEGF, PD-EGF, and prostaglandins. Some angiogenic factors may act directly on endothelial cells, whereas others may have an effect through an intermediary cell (e.g. macrophages or mast cells). Other models postulate that angiogenesis is caused by the loss of an angiogenesis inhibitors or that angiogenesis factor may be regulated by a tumor suppressor gene.

Angiogenesis constitutes an ideal target for chemotherapy. It occurs early in carcinogenesis, is essential for invasion and metastasis, appears to be regulated by extracellular signals and membrane receptors, which are excellent targets for chemotherapy, and does not occur in adult normal tissues minimizing the risk of side effects.

ACKNOWLEDGEMENT

I wish to thank Yolanda Valderrama and Carrie McKinley for their assistance in the preparation of this manuscript, John Riley for the art work and Kevin Flynn for editorial corrections. This paper was supported by DHS grants CA 42157, CA 53123, and CA 57596.

References

Arnold, F., and West, D.C. Angiogenesis in wound healing. Pharm. Ther. 52:407-422, 1991.

Ausprunk, D. Tumor angiogenesis. In: Houck, J.L. (ed) Chemical messengers of the inflammatory process. Elsevier Amsterdam, pp. 317-361, 1979.

Ausprunk, D.M., and Folkman, J. Migration and proliferation of endothelial cells in preformed and newly formed blood vessels during angiogenesis. Microvasc. Res. 14:53-65, 1977.

Balmain, A. and Brown, K. Oncogene activation in chemical carcinogenesis. Adv. Cancer Res. 51:147-182, 1988.

Bishop, J.M. Molecular themes in oncogenesis. Cell. 64:235-248, 1991.

Blood, C.H., and Zetter, B.R. Tumor interactions with the vasculature: Angiogenesis and tumor metastasis. Biochem. Biophys. Acta. 1032:89-118, 1990.

Bosari, S., Lee, A.K.C., DeLellis, R.A., Wiley, B.D., Heatley, G.J., and Silverman, M.L. Microvessel quantitation and prognosis in invasive breast carcinoma. Hum. Pathol. 23:755-761, 1992.

Brem, H. and Folkman, J. Inhibition of tumor angiogenesis mediated by cartilage. J. Exp. Med. 141:427-439, 1975.

Brem, S.S., Zagzag, D., Tsanaclis, A.M.C., Gately, S., Elkouby, M.-P., and Brein, S.E. Inhibition of angiogenesis and tumor growth in the brain. Am. J. Pathol. 137:1121-1142, 1990.

Conti, C.J. and Klein-Szanto, A.J.P. Biochemistry and Molecular Biology of Carcinomas of the Epidermis. In: Aspects of the Biochemistry and Molecular Biology of Tumors. T.G. Pretlow and T.P. Pretlow, eds., Academic Press, New York, in press.

D'Amore, P.A., and Klagsbrun, M. Angiogenesis factors and mechanisms In: Sirica, A.E. (ed) Pathobiology of neoplasia. Plenum Press New York, pp. 513-531. 1989.

DiGiovanni, J.. Multistage carcinogenesis in mouse skin. Pharmacol Ther. 54: 63-128, 1992.

Ehrmann, R.L. Knoth, M. Choriocarcinoma. Transfilter stimulation of vasoproliferation in the hamster cheek pouch. Studied by light and electron microscopy. J. Natl. Cancer. Inst. Dec. 41(6). P 1329-41, 1968..

Eisenstein, R., Sargente, N., Soble, L., Miller, A., Kuettner, K.E. The resistance of certain tissues to invasion: penetrability of explanted tissues by vascularized mesenchyme. Am. J. Pathol. 73:765-774, 1973.

Folkman, J. Langer R., Linhardt, R.J., Haudenschild, C., and Taylor, S. Angiogenesis inhibition and tumor regression caused by heparin or a heparin fragment in the presence of cortisone. Science 221:719, 1983.

Folkman, J. What is the evidence that tumors are angiogenesis dependent? J. Natl Cancer Inst. 82:4-6, 1990.

Folkman, J., Watson, K., Ingver, D., Hanahan, D. Induction of angiogenesis during the transition from hyperplasia to neoplasia. Nature 339:58-61, 1989.

Gagliardi, A., Hadd, H., and Collins, D.C. Inhibition of antiogenesis by suramin. Cancer Res. 52:5072-5075, 1992.

Good, D.J., Polverini, P.J., Rastinejad, F., Le Beau, M.M., Lemmons, R.S., Frazier, W.A., and Bouck, N.P. A tumor suppressor-dependent inhibitor of angiogenesis is immunologically and functionally indistinguishable from a fragment of thrombospondin. Proc. Natl. Acad. Sci. USA. 87:6624-6628, 1990.

Greenblatt, M., and Shubik, P. J. Natl. Cancer. Inst. Tumor angiogenesis: transfilter diffusion studies in the hamster by the transparent chamber technique. 41:111-124, 1968.

Henson, D.E, and Albores-Saavedra. Pathology of incipiet nesplasia. W.B. Saunders New York, 1993, in press.

Klagsbrun, M., D'Amore, P.A. Regulators of angiogenesis. Annu. Rev. Physiol. 53:217-239, 1991.

Knudson, A.G. Genetics of human cancer. Annu. Rev. Genet. 20:231-251, 1986.

Langer, et al. Control of tumor growth in animals by infusion of an angiogenesis inhibitorProc. Natl. Acad. Sci. U.S.A. 77:4431-4435, 1980.

Liotta, L., Kleinerman, J., Saidel, G. Quantitative relationships of intravascular tumors cells, tumor vessels, and pulmonary metastases following tumor implantation. Cancer Res. 34:997-1004, 1974.

Macchiarini, P., Fontaini, G., Hardin, M.J., Squartini, F., Angeletti, C.A. Relation of neovascularisation to metastasis of non-small-cell lung cancer. Lancet 340:145-146, 1992.

Marshall, C.J. Tumor suppressor genes. Cell 64:313-326, 1991.

Missirlis, E., Karakiulakis, G., Maragoudakis, M.E. Antitumor effect of GPA1734 in rat walker 256 carcinoma. Invest. New Drugs 8:145-147, 1990.

Montesano, R. Regulation of angiogenesis in vitro. Eur. J. Clin. Invest. 22:504-515, 1992.

Moroco, J.R., Solt, D.B., and Polverini, P.J. Sequential loss of suppressor genes for three specific functions during *in vivo* carcinogenesis. Lab. Invest. 63:298, 1990.

Murray, J.B., Allison, K., Sudhalter, J., Langer, R. Purification and partial amino acid sequence of a bovine cartilage-derived collagenase inhibitor. J. Biol. Chem. 261:4154-4159, 1986.

Nowell, P.C. Mechanisms of tumor progression. Cancer Res. 46:2203-2207, 1986.

Peraino, C., and Jones, C.A. The multistage concept of carcinogenesis In: Pathobiology of neoplasia. Plenum Press New York, pp. 131-148, 1989.

Polverini, P.J., Solt, D.B. Expression of the angiogenic phenotype by a subpopulation of keratinocytes derived from 7,12-dimethylbenz[a]anthracene-initiated hamster buccal pouch epithelium. Carcinogenesis 9:1176, 1988.

Quintanilla, M., Brown, K., Ramsden, M., and Balmain, A. Carcinogen-specific mutation and amplification of Ha-*ras* during mouse skin carcinogenesis. Nature 322:78-80, 1986.

Rastinejad, F., Polverini, P.J., Bouck, N.P. Regulation of the activity of a new inhibitor of angiogenesis by a cancer suppressor gene. Cell 56:345, 1989.

Rissau, W. Angiogenic growth factors. Prog. Growth Factor Res. 2:71-79, 1990.

Sager, R. Genetic suppression of tumor formation: A new frontier in caner research. Cancer Res. 46:1573-1580, 1986.

Sakamoto, N., Tanaka, N.G., Tohgo, A., Osada, Y., and Ogawa H. Inhibitory effect of heparin plus cortisone acetate on endothelial cell growth both in culture and in tumor masses. J. Natl. Cancer Inst. 78:585, 1987.

Sirica, A.E. Preneoplasia and precancerous lesions. In: Sirica, A.E. (ed) Pathobiology of neoplasia. Plenum Press New York, pp. 199-215, 1989.

Slaga, T.J. Cellular and molecular mechanism involved in multistage skin carcinogenesis. In: Skin tumors: experimental and clinical aspects, Conti, C.J., Slaga, T.J., and A.J.P. Klein-Szanto (eds.) Raven Press, New York. pp. 1-18, 1989.

Stanbridge, Eric, J. Human tumor suppressor genes. Annu. Rev. Genet. 24:615-657, 1990.

Sukumar, S., Notario, V., Martin-Zanca, D., Barbacid, M. Induction of mammary carcinogenesis by nitroso-methylureas involves malignant activation of H-ras locus by single point mutation. Nature 306:658-661, 1983.

Taylor, C.M., and Weiss, J.B. Partial purification of a 5.7k glycoprotein from bovine vitreous which inhibits both angiogenesis and collagenase activity. Biochem. Biophys. Res. Commun. pp. 911-6, 1985.

Vogelstein, B., Fearon, E.R., Hamilton, S.R., Kern, S.E., Preisinger, B.A., Leppert, M., Nakamura, Y., White, R., Smiths, A.M.M., and Bos, J.L. Genetic alterations during colorectal tumor development. N. Engl. J. Med. 319:525-532, 1988.

Weidner, N., Semple, J., Welch, W., and Folkman, J. Tumor angiogenesis and metastasis-correlation in invasive breast carcinoma. The N. Engl. J. .Med.. 324:1-8, 1991.

Weinberg, R.A. Oncogenes, antioncogenes, and the molecular bases of multistep carcinogenesis. Cancer Res. 49:3713-3721, 1989.

Williams, G.A., Eisenstein, R., Schumacher, B., Hsiao, K.C., Grant, D. Inhibitor of vascular endothelial cell growth in the lens. Am. J. Ophthalmol. 97:366-371, 1984.

Yuspa, S.H., and Poirer, M.C. Chemical carcinogenesis from animal models to molecular models in one decade. Adv. Cancer Res. 50:25-70, 1988.

Ziche, M., Ruggiero, M., Pasquali, F., Chiarugi, V.P. Effects of cortisone with and without heparin on angiogenesis induced by prostaglandin E1 and by S180 cells and on growth of murine transplantable tumors. Inst. J. Cancer 35:549-552, 1985.

CYTOKINE REGULATION OF TUMOR-ASSOCIATED MACROPHAGES: THERAPEUTIC IMPLICATIONS

Alberto Mantovani, Barbara Bottazzi, Silvano Sozzani, Giuseppe Peri, Paola Allavena, Cecilia Garlanda, Annunciata Vecchi and Francesco Colotta
Istituto di Ricerche Farmacologiche "Mario Negri"
Via Eritrea, 62, 20157 Milano, Italy

Macrophages are a major component of the lymphoreticular infiltrate of tumors. Tumor-derived cytokines play an important role in the regulation of the recruitment and function of tumor-associated macrophages (TAM). TAM have pleiotropic functions that influence various aspects of the immunobiology of neoplastic tissues including vascularization, stroma formation and dissolution, and growth rate. These cells, strategically located at the interface between tumor and host, have the potential to destroy neoplastic tissues and hence are a target for therapeutic intervention. Recent clinical results in ovarian cancer encourage efforts along this line.

INTRODUCTION

Macrophages are a major component of the lymphoreticular infiltrate of rodent and human tumors (Mantovani et al 1992a). Since these cells are situated at the very interface between tumor and host, they may represent a strategically located target for therapeutic intervention. Interest in these cells is stimulated by the knowledge that macrophages have the potential to kill neoplastic cells including drug-resistant variants surviving conventional chemotherapy.

Tumor-associated macrophages (TAM) derive from circulating monocytic precursors. Tumor-derived chemotactic factors (TDCF) have been identified.

In addition to releasing chemoattractants various human tumor

NATO ASI Series, Vol. H 75
Cancer Therapy
Edited by N. D'Alessandro, E. Mihich, L. Rausa, H. Tapiero, and T. R. Tritton

lines release an inhibitor of chemotaxis antigenically related to the transmembrane protein P15E of murine retroviruses (Snyderman and Cianciolo 1984; Wang et al 1986). Hence the regulation of monocyte infiltration in tumors is complex and may involve a balance of factors with opposing influences on leukocyte migration.

The functional properties of macrophages infiltrating murine and human metastatic tumors have been characterized in an effort to obtain indications as to the role played by these cells in the immunobiology of neoplastic tissues (Mantovani et al 1992a). This analysis has indicated how TAM can contribute to important aspects of tumor tissue biology, such as fibrin deposition and angiogenesis.

Moreover TAM in certain tumors are a source of growth factors which actually provide the optimal conditions for tumor growth. More in general, this type of analysis has shown how TAM, within the mononuclear phagocyte system, represent a population with peculiar phenotypic and functional properties. Here, we will summarize recent work on molecules involved in the regulation of macrophage infiltration and function in neoplastic tissues.

MCP-1, A CYTOKINE INVOLVED IN THE REGULATION OF TAM

TAM originate from circulating monocytic precursors. Tumors vary widely in the size of their macrophage infiltrate although this is relatively stable for each neoplasm. For many murine tumors, the number of TAM is not affected by transplantation of the tumor into mice with defective T-cell-mediated immunity (thymectomized, nude or ultraviolet-irradiated). This suggests that, in many tumors, specific immunity is not an important determinant of macrophage infiltration, and that factors inherent to the tumor itself regulate macrophage levels in neoplastic tissues (for review, see Mantovani et al 1992a).

This consideration prompted us to search for monocyte chemoattractants released by tumor cells (Bottazzi et al 1983). We have identified a cytokine of about 12kDa, called tumor-derived chemotactic factor (TDCF), which is chemotactic for monocytes and is released by various murine and human tumor cells. Recently, the sequence and structure of this cytokine, which we now refer to as monocyte chemotactic protein-1 (MCP-1) have been determined (Furutani et al 1989, Yoshimura et al 1989, Van Damme et al 1989, Bottazzi et al 1990, Matsushima et al 1989). MCP-1 is a 76-amino-acid mature polypeptide with one N-linked glycosylation site. The gene encoding MCP-1 is identical to JE, a gene identified in activated fibroblasts, and is homologous to genes expressed in activated lymphocytes and mesenchimal cells (for a review see Oppenheim et al 1991). The structural hallmarks of these cytokines are four conserved cysteines, the first two of which are in tandem, and which are probably important for the three-dimensional structure. The same four-cysteine motif is shared by cytokines of the platelet factor (PF)4 family, except that the first two cysteines are interrupted by an intervening amino acid. The PF4 family includes interleukin (IL)-8 and melanocyte growth-stimulating activity (MGSA/gro), which are chemotactic for neutrophils. Thus, MCP-1 belongs to an emerging superfamily of cytokines, some of which are involved in the regulation of leukocyte recruitment and activation. The term intercrines and, more recently, chemokines, has been proposed for these mediators. MCP-1 is produced by a variety of cell types, including lymphocytes, mononuclear phagocytes and fibroblasts. Particularly significant in the context of monocyte recruitment is the fact that it is produced by vascular smooth muscle cells and endothelial cells (Valente et al 1984, Sica et al 1990).

As expected from its _in vitro_ chemotactic activity, injection of MCP-1 _in vivo_ induces extravasation of monocytes (Zachariae et al 1990). There is also evidence that it has an important role in the regulation of TAM levels _in vivo_. In a series of murine tumors, in tumor biopsies from ovarian carcinomas, and in human tumor variants transplanted into nude mice, a significant, though far from absolute, correlation was found between the number of TAM and the MCP activity produced by the tumor cells (Bottazzi et al 1983; Walter et al 1991). Finally, MCP-1 gene transfer resulted in higher levels of TAM in a

murine melanoma (Bottazzi et al 1992).
The metastatic capacity of clones expressing the human MCP-1 gene was studied by injecting tumor cells i.v. MCP-1 expressing clones were more metastatic than control cells in terms of lung involvement (number or weight) and of the occurrence of extrapulmonary lesions (Table 1). This observation is in line with a higher tumorigenicity at low tumor inocula in spite of a slower _in vivo_ growth rate (Bottazzi et al 1992). Mononuclear phagocytes recruited by MCP-1 may help the initial implantation and outgrowth of the small number of cells that ultimately give rise to metastasis.

PARACRINE REGULATION OF TAM SURVIVAL AND PROLIFERATION

The mechanisms involved in the maintenance of constant levels of macrophages in growing tumors are complex and involve various factors. It has been reported that TAM have increased proliferative capacity (e.g. Evans and Cullen 1984, Mahoney and Heppner 1987, Bottazzi et al 1990) and _in situ_ proliferation may contribute to the macrophage content of tumor tissues.
The proliferative capacity of TAM from two murine sarcomas was investigated by cytofluorography: the frequency of cells in the S phase of the cell cycle was 7-11% for TAM and 1-2% for resident or elicited macrophages. The proliferation and differentiation of mononuclear phagocytes is regulated by the growth factor M-CSF which is also active on differentiated macrophages. The c-fms proto-oncogene encodes for a transmembrane glycoprotein probably identical to the M-CSF receptor. We therefore examined c-fms expression in TAM. TAM showed levels of c-fms mRNA higher than peritoneal exudate macrophages (PEM). Having established that TAM express c-fms at levels higher than PEM, it was of interest to investigate M-CSF expression in TAM and sarcoma cells. TAM did not express appreciable levels of M-CSF mRNA while tumor cells from both sarcomas used in this study showed high levels of M-CSF transcripts. Moreover, supernatants from 2 fibrosarcomas have M-CSF activity on bone marrow cells and induced proliferation

of peritoneal exudate and bone marrow-derived macrophages. These activities were blocked by anti-M-CSF antibody.

These observations outline the existence of a paracrine circuit in the regulation of TAM survival and proliferation, involving M-CSF, secreted by sarcoma cells and acting on c-fms expressing TAM.

It is of interest that TAM from human Kaposi's sarcoma have been shown recently to have proliferative activity (C. Parravicini, unpublished).

ANTITUMOR POTENTIAL OF MACROPHAGES

Appropriately activated mononuclear phagocytes can kill tumor cells _in vitro_ and elicit tumor-destructive reactions _in vivo_. The latter involve vascular responses that are induced by cytokines (TNF and IL-1) released by monocytes (Mantovani et al 1992b). Macrophages can kill tumor cells through a variety of mediators, including soluble or membrane-associated cytokines (IL-1 and TNF), and reactive oxygen or nitrogen intermediates and probably other, as yet undefined, cytotoxic pathways. In most experimental systems, macrophage cytotoxicity involves a close interaction with tumor target cells. We have recently found that antibodies to leukocyte integrins (lymphocyte function-associated antigen (LFA)-1 in particular) inhibit human monocyte cytotoxicity including that exerted on targets that are sensitive to TNF (Bernasconi et al 1991, Jonjic et al 1992). These findings are compatible with a model in which contact between monocyte surface integrins and undefined ligands on tumor target cells allows local delivery of cytotoxic molecules.

Further evidence for the importance of the CD18-dependent adhesion pathway was recently obtained by a gene transfer approach. Transfer of the ICAM-1 gene in a "low ICAM-1" melanoma clone (2/21) relatively resistant to monocyte cytotoxicity resulted in augmented levels of killing (Jonjic et al, unpublished).

IMMUNOTHERAPY IN OVARIAN CARCINOMA

While the results obtained in rodent tumors and *in vitro* suggest that activated macrophages indeed have the potential to eliminate at least small numbers of tumor cells surviving after cytoreductive therapy, evidence for antitumor activity of macrophage activators in humans is limited. In the context of a long standing interest in exploring i.p. immunotherapy in ovarian cancer (e.g. Allavena et al 1990) attention was recently focused on IFNγ, a prototypic macrophage activating cytokine. In a large cooperative study with 98 patients, i.p. IFNγ showed definite antitumor activity in subjects with minimal residual disease resistant to chemotherapy (Allavena et al 1992, Pujade-Lauraine et al 1990). It is of interest that when IFNγ was administered systematically, no modulation of in situ effector cells and no clinical response were observed (Colombo et al., unpublished).

CONCLUDING REMARKS

Phagocytes infiltrating neoplastic tissues have peculiar membrane phenotype and functional properties. Thus TAM play a complex, ambiguous role in the regulation of primary tumor growth and metastasis (a "macrophage balance", see Mantovani et al 1992a). Yet these cells are strategically located at the very interface between tumor and host and represent a potential target for immunomodulation. A better understanding of the regulation and function of TAM may provide a less empirical basis for rational design of therapeutic approaches, as vividly illustrated by the antitumor activity of i.p. interferonγ in ovarian cancer patients with minimal residual disease resistant to chemotherapy.

Acknowledgements
This work was supported by finalized project ACRO. The generous contribution of the Italian Association for Cancer Research, Milan, Italy, is gratefully acknowledged.

Table 1

Metastatic capacity of melanoma cells after MCP-1 gene transfer[a]

Exp.	*Clone*	*MCP-1*	*Mice with met.*	*Lung colonies (total number)*	*Weight (mg ± SE)*	*Extrapulmonary lesions*
1	V14	-	7/8	20	210 ± 21	0/8
	L12	+	5/5	<300	862 ± 156	1/5
2	V14	-	7/10	16	138 ± 66	1/10
	V16	-	9/9	23	76 ± 17	0/9
	L12	+	8/8	14	296 ± 116	8/8
	L4	+	7/7	15	174 ± 80	0/7

a) The human MCP-1 gene under the control of a retroviral promoter was transfered in the B78/HI subline of the B16 melanoma. The obtainement and properties of MCP-1 expressing (L12-L4) and control (V14-V16) clones are described by Bottazzi et al (1992). 5 x 10 cells of in vitro cultured cells were injected in the tail vein and mice were autopsied on day 30 (experiment 1) or 34 (experiment 2).

b) Extrapulmonary metastases involved lymph nodes and liver.

REFERENCES

Allavena, P., Grandi, M., D'Incalci, M., Geri, O., Giuliani, F.C., and Mantovani, A., (1987). Human tumor cell lines with pleiotropic drug resistance are efficiently killed by Interleukin-2 activated killer cells and by activated monocytes. *Int. J. Cancer,* **40**: 104-107

Allavena, P., Peccatori, F., Maggioni, D., Sironi, M., Colombo, N., Lissoni, A., Galazka, A., Meiers, W., Mangioni, C., and Mantovani A., (1990). Intraperitoneal recombinant g-interferon in patients with recurrent ascitic ovarian carcinoma: modulation of cytotoxicity and cytokine production in tumor-associated effectors and major histocompatibility antigen expression on tumor cells. *Cancer Res.,* **50**: 7318-7323

Bernasconi, S., Peri G., Sironi, M., and Mantovani, A., (1991). Involvement of leukocyte (B2) integrins (CD18/CD11) in human monocyte tumoricidal activity. *Int. J. Cancer,* **49**: 267-273

Bottazzi, B., Polentarutti, N., Acero, R., Balsari, A., Boraschi, D., Ghezzi, P., Salmona, M., and Mantovani, A., (1983). Regulation of the macrophage content of neoplasms by chemoattractants. *Science* **220**: 210-212

Bottazzi, B., Colotta, F., Sica, A., Nobili, N. and Mantovani, A., (1990a). A chemoattractant expressed in human sarcoma cells (Tumor-derived chemotactic factor, TCDF) is identical to monocyte chemoattractant protein-1/monocyte chemotactic and activating factor (MCP-1/MCAF). *Int. J. Cancer,* **45**: 795-797

Bottazzi, B., Erba, E., Nobili, N., Fazioli, F., Rambaldi, A. and Mantovani, A., (1990b). A paracrine circuit in the regulation of the proliferation of macrophages infiltrating murine sarcomas. *J. Immunol.,* **144**: 2409-2412

Bottazzi, B., Walter, S., Govoni, D., Colotta F., and Mantovani, A., (1992). Monocyte chemotactic cytokine gene transfer modulates macrophage infiltration, growth and susceptibility to IL-2 therapy of a murine melanoma. *J. Immunol.,* **148**: 1280-1285

Colombo, N., Peccatori, F., Paganin, C., Bini, S., Brandely, M., Mangioni, C., Mantovani, A., and Allavena, P. Anti-tumor and immunomodulatory activity of intraperitoneal administration of IFN g in ovarian carcinoma patients with minimal residual tumor after chemotherapy, *Int. J. Cancer,* 1992 - *in press.*

Evans, R., and Cullen, R.T., (1984). In situ proliferation of intratumor macrophages. *Journal Leukocyte Biology* **35**: 561-572

Furutani, Y., Nomura, H., Notake, M., Oyamada, Y., Fukui, T., Yamada, M., Larsen, C.G., Oppenheim, J.J., and Matsushima, K., (1989). Cloning and sequencing of the cDNA for human monocyte chemotactic and activating factor (MCAF). *Biochem. Biophys. Res. Commun.* **159**: 248-255

Jonjic, N., Jìlek, P., Bernasconi, S., Peri, G., Martin-Padura, I., Cenzuales, S., Dejana, E., and Mantovani, A., (1992). Molecules involved in the adhesion and cytotoxicity of activated monocytes on endothelial cells. *J. Immunol.,* **148**: 2080-2083

Mahoney, K.H., and Heppner, G.H., (1987). FACS analysis of tumor associated macrophage replication: differences between metastatic and nonmetastatic murine mammary tumors. *Journal Leukocyte Biology* **41**: 205-211

Mantovani, A., Bottazzi, B., Colotta, F., Sozzani, S., and Ruco, L., (1992a). Origin and function of tumor-associated macrophages. *Immunology Today* **13**: 265-299

Mantovani, A., Bussolino, F., and Dejana, E., (1992b). Cytokine regulation of endothelial cell function. *Faseb J.* **6**: 2591-2599

Matsushima, K., Larsen, C.G., DuBois, G.C., and Oppenheim, J.J., (1989). Purification and characterization of a novel monocyte chemotactic and activating factor produced by a human myelomonocytic cell line. *J. Exp. Med.* **169**: 1485-1490

Oppenheim, J.J., Zachariae, C.O.C., Mukaida, N., and Matsushima, K., (1991). Properties of the novel proinflammatory supergene "Intercrine" cytokine family. *Annual Review of Immunol.,* **9**: 617-648

Pujade-Lauraine, E., Colombo, N., Namer, N., Fumoleau, P., Monnier, A., Nooy, M.A., Falkson, G., Mignot, L., Bugat, R., Oliveira, C.M.D., Mousseau, M., Netter, G., Oberling, F., Coiffier, B., and Brandely, M., (1990). Intraperitoneal human r-IFN gamma in patients with residual ovarian carcinoma (OC) at second look laparotomy (SLL). *ASCO,* **9**: 156 abs., 111

Sica, A., Wang, J.M., Colotta, F., Dejana, E., Mantovani, A., Oppenheim, J.J., Larsen, C.G., Zachariae, C.O., and Matsushima, K. (1990). Monocyte chemotactic and activating factor gene expression induced in endothelial cells by IL-1 and tumor necrosis factor. *J. Immunol.,* **144**: 3034-3038

Snyderman, R., and Cianciolo, G.J., (1984). Immunosuppressive activity of the retroviral envelope protein P15E and its possible relationship to neoplasia. *Immunology Today,* **5**: 240-244

Valente, A.J., Fowler, S.R., Sprague, E.A., Kelley, J.L., Suenram, A.C. and Schwartz, C.J., (1984). Initial characterization of a peripheral blood mononuclear cell chemoattractant derived from cultured arterial smooth muscle cells. *Am. J. Pathol.,* **117**: 479-485

Van Damme, J., Decock, B., Lenaerts, J.P., Conings, R., Bertini R., Mantovani, A. and Billiau, A., (1989). Identification by sequence analysis of chemotactic factors for monocytes produced by normal and transformed cells stimulated with virus, double stranded RNA or IL-1. *Europ. J. Immunol.,* **19**: 2367-2373

Walter, S., Bottazzi, B., Govoni, D., Colotta, F., and Mantovani, A., (1991). Macrophage infiltration and growth of sarcoma clones expressing different amounts of monocyte chemotactic protein/JE. *Int. J. Cancer,* **49**: 431-435

Wang, J.M., Cianciolo, G.J., Snyderman, R., and Mantovani, A., (1986). Coexistence of a chemotactic factor and a retroviral P15E-related chemotaxis inhibitor in human tumor cell culture supernatants. *J. Immunol.* **137**: 2726-2732

Yoshimura, T., Yuhki, N., Moore, S.K., Appella, E., Lerman, M.I., and Leonard, E.J., (1989). Human monocyte chemoattractant protein-1 (MCP-1). Full-lenght cDNA cloning, expression in mitogen-stimulation blood mononuclear leukocytes , and sequence similarity to mouse competence gene JE. *FEBS Letters* **244**: 487-493

Zachariae, C.O.C., Anderson, A.O., Thompson, H.L., Appella, E., Mantovani, A., Oppenheim, J.J. and Matsushima, K., (1990). Properties of monocyte chemotactic and activating factor (MCAF) purified from a human fibrosarcoma cell line. *J. Exp. Med.*, **171**: 2177-2182

THE MECHANISM OF LECTIN-MEDIATED ADHESION OF HUMAN OVARIAN CARCINOMA CELLS

Dr. R.J. Bernacki[1]
Department of Experimental Therapeutics
Roswell Park Cancer Institute
Buffalo, New York USA 14263

Introduction

Tumor invasion, dissemination and the subsequent formation of metastatic foci derive from alterations in tumor cell adhesiveness. Lectins and other intercellular adhesion molecules (ICAMs) are involved in establishing adhesion between tumor cells, vascular endothelial cells and the basement membrane. Galaptin, an endogenous lectin with affinity for galactose containing oligosaccharides, is present in the extracellular matrix (ECM) and participates in the binding of A121 human ovarian carcinoma cells to the ECM. The specificity for galactose was demonstrated by inhibiting binding with lactose and by pretreating A121 cells with ß-galactosidase. We have now carried out studies to identify the cellular glycoprotein(s) that are responsible for binding A121 cells to galaptin. Our results indicate that the galaptin receptor on the ovarian tumor cell surface is a highly glycosylated, lysosome-associated membrane protein (lamp). A series of membrane sugar analogs has been designed and synthesized as potential modulators of the galaptin receptor. One analog, 2-acetamido-1,4,6-tri-O-acetyl-3-deoxy-3-fluoro-glucosamine (CD 89029), was found to specifically inhibit glycoprotein biosynthesis. As a result of this inhibition, tumor cells treated with CD 89029 decreased their adhesion to the ECM. These studies show that tumor cell adhesion is mediated by specific endogenous lectins and their cell surface glycoconjugate receptors (i.e. lamp), and that appropriate sugar analogs can serve as potential inhibitors of metastatic spread.

[1]R.J. Bernacki, D.M. Skrincosky, H.J. Allen, K.L. Matta, M. Sharma, A.F. Haag and B. Woynarowska. Departments of Experimental Therapeutics (RJB, DMS, MS, AFH, BW) and Gynecologic Oncology (HJA, KLM), Roswell Park Cancer Institute, Buffalo, New York USA 14263

NATO ASI Series, Vol. H 75
Cancer Therapy
Edited by N. D'Alessandro, E. Mihich, L. Rausa, H. Tapiero, and T. R. Tritton

Nature of Ovarian Cancer Metastasis

Metastasis involves the dissemination of tumor cells from the site of origin to distal locations. In ovarian cancer, tumor cells often invade the peritoneal cavity establishing metastatic foci on the mesothelial linings of various internal organs and structures. Growth of these secondary tumors often results in the impairment of visceral organ function (e.g. bowel obstruction), and sloughing of tumor cells from these sites can block lymphatic drainage, causing an accumulation of large amounts of ascitic fluid.

Development of an *In Vitro* Ovarian Tumor Model

In order to study the various steps involved in this metastatic cascade, we developed an *in vitro* model illustrated schematically in Fig. 1.

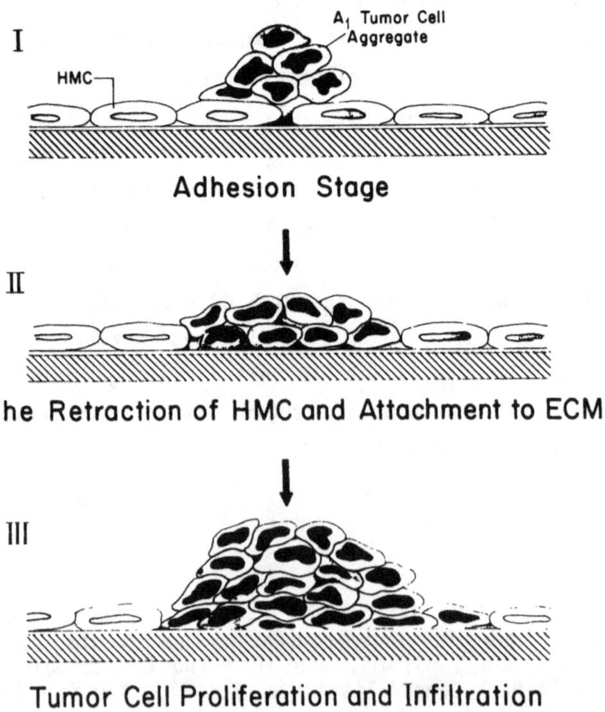

Fig. 1 *In vitro* ovarian tumor model system. Human mesothelial cells (HMC) are grown on extracellular matrix deposited by bovine corneal endothelial cells. Human ovarian carcinoma cells (e.g. A1 cell line) adhere and proliferate on this "artificial" mesothelium.

Corneal endothelial cells, derived from fresh bovine corneas, grow in primary culture when basic fibroblast growth hormone (bFGF) or epidermal growth factor (EGF) are supplied (Gospodarowicz et al., 1977). Upon reaching confluence, these cells secrete a basement membrane like substance termed "extracellular matrix" (ECM). That matrix, which contains fibronectin, laminin, collagen IV and various proteoglycans, serves as an excellent substrate for the adhesion and subsequent growth of normal human mesothelial cells (HMC).

We observed that fresh human ovarian carcinoma cells or cell lines developed therefrom [e.g. A1 or A121 (Crickard et al. 1989)] adhere to and proliferate on "artificial" mesothelium, forming colonies. The tumor cells displayed a hundred fold greater affinity for the ECM as compared to the "luminal" mesothelial cell surface. Scanning electron microscopy showed that the tumor cells seek out areas of exposed ECM and cause retraction of the mesothelial cell monolayer providing a larger surface for subsequent tumor cell adhesion and growth (Niedbala et al. 1985).

Tumor Cell Degradation of EOM

Adhesion and growth of human ovarian tumor cells on ECM results in the degradation of ECM (Niedbala et al. 1987). This event, which facilitates invasion and metastasis, is likely a consequence of the release of lysosomal hydrolases, including proteases and glycosidases (Woynarowska et al. 1989). These findings prompted us to search for agents capable of inhibiting hydrolase activity and decreasing the tumor cell mediated degradation of ECM.

Several glycosidase and protease inhibitors demonstrated "antimetastatic" activity. Thus, 2-acetamido-2-deoxy 1,5-gluconolactone and 2-acetamido-1,5-imino-1,2,5-trideoxy-D-glucitol inhibited A121 ß-N-acetyl-glucosaminidase (ß-NAG) activity in A121 ovarian carcinoma cells and prevented the cell mediated degradation of the ECM (Woynarowska et al. 1992). Similarly, the protease inhibitors leupeptin and pepstatin A decreased the tumor cell-mediated degradation of ECM and metastasis of solid tumors in mice (Leto et al. 1990). These findings suggest that interference with tumor cell-mediated degradation of the basement membrane results in decreased tumor invasiveness and metatasis.

Galaptin Mediates Adhesion of Human Ovarian Carcinoma Cells to the ECM

In an attempt to elucidate the mechanism(s) of ovarian tumor cell adhesion to the ECM, we investigated the possibility of lectin involvement. Previous studies have shown that the leukocyte adhesion to endothelium, which occurs during inflammation, depends on the endothelial cells recognizing leukocyte surface glycoconjugates via a specific group of lectins, termed selectins. Endothelial cell surface selectins or ELAMs (endothelial-leukocyte adhesion molecules) become upregulated during inflammation. One putative leukocyte surface selectin receptor appears to be the sialylated-Lewis X antigen, S-Le^X (Phillips *et al*. 1990; Polley *et al*. 1991) and selectin interaction with S-Le^X is thought to result in the deceleration or "rolling" of leukocytes in capillaries, leading endothelial cells to secrete factors which induce the expression of CD18 integrins on leukocyte surfaces. This modulation leads to the "permanent" attachment of leukocytes to endothelial cells, which facilitates invasion of the capillary endothelial basement membrane by the leukocytes (Springer, 1990; Butcher, 1991).

Similarities exist between leukocyte adherence to endothelial cells and their subsequent migration through the basement membrane and metastatic tumor cell adhesion and invasion. Using antibodies to human splenic galaptin, a ß-galactoside specific lectin, we showed that galaptin was observed to be associated with both A121 ovarian carcinoma cells as well as the ECM. Addition of exogenous galaptin blocked A121 cell adhesion to ECM (Allen *et al*. 1990). We, therefore, investigated whether purified galaptin can affect cellular adhesion, and whether a glucosamine analog (CD 89029) is capable of modulating cell surface glycoconjugate biosynthesis and cellular adherence.

Effect of Galaptin on A121 Cell Adhesion

As shown in Fig. 2, coating culture dishes with purified galaptin increased A121 cell adhesion. At levels of galaptin above 5 µg/cm^2 adhesion began to plateau. Polymerization of galaptin with 0.1% glutaraldehyde dramatically improved A121 cell adhesion. Adhesion was enhanced by pretreatment of A121 cells with *Clostridium perfringens* neuraminidase but was decreased following treatment with Jack Bean ß-galactosidase or by the addition of 50 mM lactose . These findings strongly suggest that adhesion of A121 cells to polymerized-galaptin coated tissue culture plates is dependent on the presence of intact cell surface galactoside residues (Skrincosky *et al*., 1993).

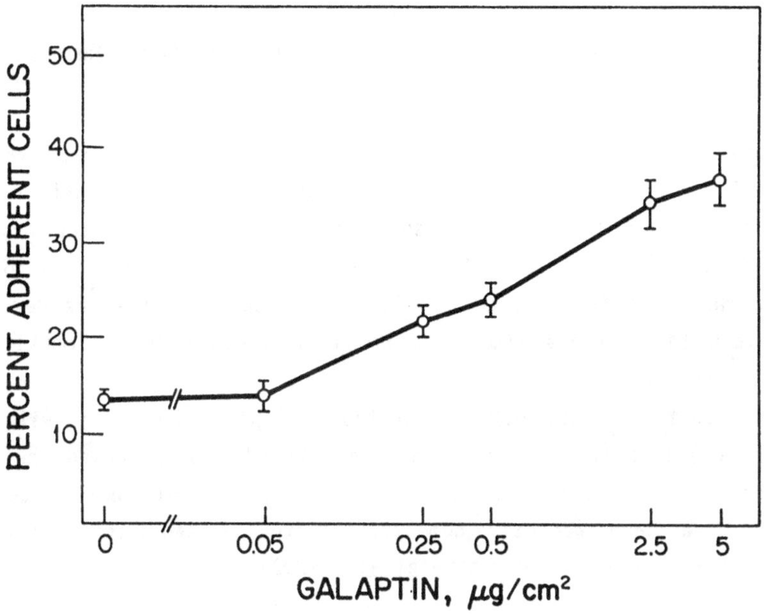

Fig. 2 Adhesion of A121 human ovarian carcinoma cells to culture plates as a function of galaptin concentration. Plates were coated with increasing concentrations of galaptin and A121 cells were added. Adhesion was measured after a 1h incubation performed at 37°. Data points are the means ± S.D. of quadruplicate samples.

The A121 Human Ovarian Carcinoma Cell Surface Galaptin-Receptor

In order to obtain some information on the nature of the galaptin receptor of A121 ovarian carcinoma cells, ^{125}I-labelled polymerized-galaptin (PG) was applied to Western blots of A121 whole cell homogenates.

The resulting autoradiogram of the Western blot indicated that a glycoprotein of 110 kDa was recognized; this binding was prevented by the addition of lactose. Purification of the A121 galaptin receptor was achieved using polymerized-galaptin bound to Sepharose. Pretreatment of PG-Sepharose affinity purified galaptin receptor with N-glycanase (to remove asparagine-linked oligosaccharides) abolished binding, confirming the carbohydrate specificity of galaptin binding (Skrincosky et al., 1993).

A121 Cellular Galaptin Receptor and Lysosomal Associated Membrane Protein (LAMP)

Earlier studies with A121 ovarian carcinoma cells showed that these metastatic tumor cells secrete high levels of lysosomal hydrolases including ß-N-acetylglucosaminidase, which participate in the degradation of the ECM. Based on these findings, we investigated possible similarities between the galaptin receptor and lysosomal membrane proteins demonstrated to contain high levels (50%) of polylactosaminylglycan (Youakim et al. 1989; Carlsson and Fukuda, 1990).

Treatment of SDS-PAGE separated galaptin receptor with rabbit polyclonal antibodies raised against lamp-1 and lamp-2 showed the galaptin receptor to be similar, if not identical, to lamp-1 and/or lamp-2. In addition, we have recently demonstrated the presence of lamps on the surface of A121 cells (Skrincosky et al., 1993)

Modulation of Tumor Cell Surface Glycoconjugate Using a Fluorinated Glucosamine Analog

Changing cell surface proteins that affect adhesion and metastasis can lead to the modification of tumor spread. Several carbohydrate analogs have been synthesized (Sharma et al., 1988) that were designed as oligosaccharide chain terminators or as inhibitors of enzymes involved in glycoprotein biosynthesis. 2-Acetamido-1,4,6-tri-O-acetyl-3-deoxy-3-fluoroglucosamine specifically inhibited glycoprotein biosynthesis decreasing cellular adhesion of A121 cells to the ECM (Fig. 3).

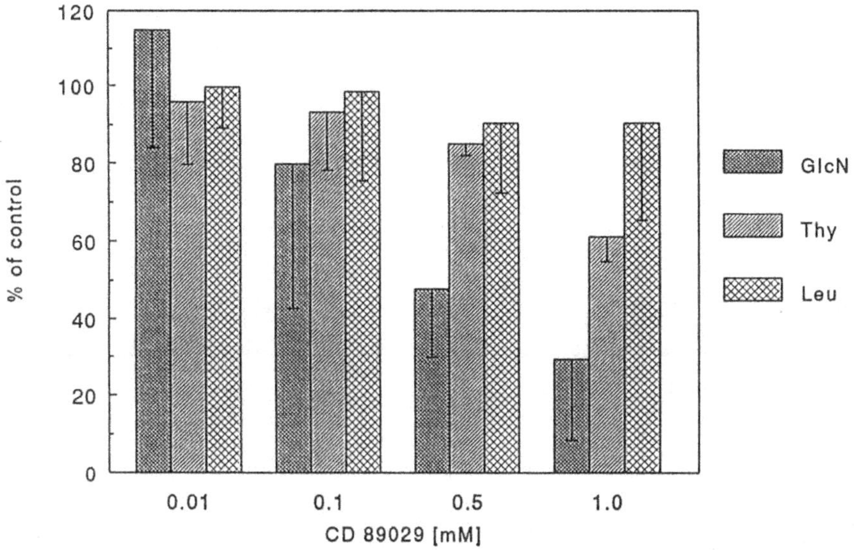

Fig. 3. Effect of 2-acetamido-1,4,6-tri-*O*-acetyl-3-deoxy-3-fluoro-glucosamine on glycoprotein, DNA and protein synthesis in A121 ovarian cancer cells. D-[6-^{3}H]-Glucosamine, [methyl-^{3}H]-thymidine, and L-[4,5-^{3}H(N)]-leucine incorporation into A121 cells was measured. The results are the mean ± S.D. and expressed as the percentage of incorporation compared to the control cells.

Conclusions

A model system has been developed that allows the mechanisms of ovarian carcinoma metastasis to be studied *in vitro*. The carcinoma cells bind preferentially to extracellular matrix (ECM) as opposed to the surface formed by normal human mesothelial cells. Various components of the ECM serve as receptors for tumor cell adhesion including galaptin, a 14.5 kDa endogenous lectin, with specificity for cell surface galactosides.

Galaptin receptor has been shown to be present on the A121 cell surface. It is a glycoprotein of 110 kDa with homology to lysosome associated membrane protein (lamp). We have also demonstrated that a fluorinated glucosamine analog, which decreases glycoconjugate biosynthesis, has the ability to depress A121 cellular adhesion to the ECM. These findings suggest that tumor cell membrane glycoconjugate- and more specifically galaptin-mediated adhesion may be a suitable target for therapeutic exploitation.

Acknowledgements

These studies were partially supported by NCI Grants CA13038, CA42898 (RJB), CA42584 (HJA) and CA35329 (KLM). We kindly thank Dr. Fukuda (La Jolla Cancer Research Foundation) for supplying antibodies to lamp and Mae Brown for typing this manuscript.

References

Allen HG, Sucato D, Woynarowska B, Gottstine S, Sharma A, Bernacki RJ (1990) Role of galaptin in ovarian carcinoma adhesion to extracellular matrix, J Cell Biochem 43, 43-57

Butcher EC (1991) Leukocyte-endothelial cell recognition: Three (or more) steps to specificity and diversity, Cell 67, 1033-1036

Carlsson SR, Fukuda M (1990) The polylactosaminoglycans of human lysosomal membrane glycoproteins lamp-1 and lamp-2, J Biol Chem 265, 20488-20495

Crickard K, Niedbala MJ, Crickard U, Yoonessi M, Sandberg AA, Okuyama K, Bernacki RJ, Satchidanand (1989) Characterization of human ovarian and endometrial carcinoma cell lines established on extracellular matrix, Gynecol Oncol 32, 163-173

Gospodarowicz D, Mescher AL, Burdwell CR (1977) Stimulation of corneal endothelial cell proliferation in vitro by fibroblast and epidermal growth factors, Exp Eye Res 25, 75-89

Leto G, Tumminello FM, Gebbia N, Woynarowska B, Bernacki RJ (1990) Antimetastatic activity of adriamycin in combination with protease inhibitors in mice, Anticancer Res 10, 265-270

Niedbala M, Crickard K, Bernacki R (1987) In vitro degradation of extracellular matrix by human ovarian tumor cells, Clin Expl Metastasis 5, 181-197

Niedbala MJ, Crickard K, Bernacki RJ (1985) Interactions of human ovarian tumor cells with human mesothelial cells grown on extracellular matrix: An in vitro model system for studying tumor cell adhesion and invasion, Exp Cell Res 160, 499-513

Phillips ML, Nudelman E, Gaeta FCA, Perez M, Singhal AK, Hakomori S-I, Paulson JC (1990) ELAM-1 mediates cell adhesion by recognition of a carbohydrate ligand, sialyl-le, Science 250, 1130-1135

Polley MJ, Phillips ML, Wayner E, Nudelman E, Singhal AK, Hakomori S-I, Paulson JC (1991) CD62 and endothelial cell-leukocyte adhesion molecule 1 (ELAM-1) recognize the same carbohydrate ligand, sialyl-Lewis x, Proc Natl Acad Sci 88, 6224-6228

Sharma M, Bernacki RJ, Korytnyk W (1988) Fluorinated analogs of cell-surface carbohydrates as potential chemotherapeutic agents. In: Taylor NF (ed) Fluorinated Carbohydrates: Chemical and Biochemical Aspects ACS Symposium Series No. 374 Chapter 11. American Chemical Society, Washington DC, pp 191-206

Skrincosky DM, Allen HJ, Bernacki RJ (1993) Galaptin mediated adhesion of human ovarian carcinoma A121 cells, Cancer Research (In Press)

Springer TA (1990) Adhesion receptors of the immune system, Nature 346, 425-434

Woynarowska B, Wikiel H, Bernacki RJ (1989) Human ovarian carcinoma ß-N-acetylglucosaminidase isoenzymes and their role in extracellular matrix degradation, Cancer Research 49, 5598-5604.

Woynarowska B, Wikiel H, Sharma M, Carpenter N, Fleet GWJ, Bernacki RJ (1992) Inhibition of human ovarian carcinoma cell- and hexosaminidase-mediated degradation of extracellular matrix by sugar analogs. Anticancer Res 12, 161-166

Youakim A, Romero PA, Yee K, Carlsson SR, Fukuda M, Herscovics A (1989) Decrease in polylactosaminoglycans associated with lysosomal membrane glycoproteins during differentiation of CaCo-2 human colonic adenocarcinoma cells, Cancer Research 49, 6889-6895

INHIBITORY EFFECT OF SURAMIN AND HEPARIN-LIKE DRUGS ON EXPERIMENTAL ANGIOGENESIS

R. Danesi[1], M. Costa[2], C. Agen[2], U. Benelli[2], M. Del Tacca[2]

[1]School of University Studies
and Doctoral Research S. Anna
Via Carducci 40
I-56127 Pisa Italy

[2]Institute of Medical Pharmacology
University of Pisa
Via Roma 55
I-56126 Pisa Italy

Introduction

In many pathological states, including tumor growth and a wide variety of non-neoplastic diseases, the disease itself is driven by persistent and often uncontrolled neovascularization (Folkman, 1992). A large number of factors derived from normal and pathologic tissues stimulate the proliferation of vascular cells *in vitro* and promote neovascularization *in vivo*. The heparin-binding growth factors, including the family of the fibroblast growth factors (FGFs), are potent inductors of angiogenesis in several experimental models (Risau, 1990). Heparin and heparan sulfate present in the side chain of cell surface proteoglycans have been shown to protect bFGF from proteolytic degradation and have been suggested to act as a storage reservoir for FGFs; in addition to this, they are required for bFGF binding to FGF receptors (Saksela et al, 1988). Suramin is a polysulfonated naphtylurea that has been shown to block the cell surface binding of various growth factors such as bFGF and platelet-derived growth factor (PDGF) (La Rocca et al, 1990), and is currently used for the treatment of tumors, including prostate cancer (Myers et al, 1992).

The present study investigates the effect of suramin, heparin, heparan sulfate and hydrocortisone on *in vitro* growth of endothelial cells and *in vivo* angiogenesis in the chick chorioallantoic membrane and in the rat cornea.

Materials and methods

Bovine aortic endothelial cell (BAEC) cultures

Cloned populations of BAEC were established from the intima of bovine aorta as described (Baldin et al, 1990). Stock cultures were grown in Dulbecco's modified Eagle's Medium (D-MEM) containing glucose (1 g/l), heat-inactivated fetal bovine serum (10%), L-glutamine (2 mM), penicillin (50 IU/ml) and streptomycin (50 μg/ml); bFGF (2 ng/ml) was added every other day. Cells were harvested with trypsin-EDTA for 1-5 min at 4 °C, resuspended in D-MEM with

NATO ASI Series, Vol. H 75
Cancer Therapy
Edited by N. D'Alessandro, E. Mihich, L. Rausa,
H. Tapiero, and T. R. Tritton

supplements and cultured at 37 °C, 5% CO_2 atmosphere. Cells were never allowed to reach confluency to avoid contact inhibition of proliferation and cells in log-phase of growth were used for experiments. Quiescent sparse endothelial cells were obtained as described (Baldin et al, 1990). Briefly, cells were seeded at low density (10^3 cells/cm^2) in 24-well plates in complete medium containing bFGF (2 ng/ml); after 72 h, to arrest cells in G_1 phase, they were washed twice with serum-free D-MEM containing Fe^{2+}-saturated transferrin (10 μg/ml) and cultures were continued in the same medium for 48 h. The effects of suramin and heparan sulfate were evaluated on BAEC in log-phase of growth and on bFGF-stimulated quiescent cells by ^{3}H-thymidine incorporation into DNA.

Chick chorioallantoic membrane (CAM) assay
Fertilized white Leghorn chick eggs were incubated for 3 days and then washed in Betadine® and 70% ethanol, cracked and shell was separated. The embryos were placed in a sterile *in vitro* system which uses plastic wrap to mimic natural egg contours (Dugan et al, 1991) and 4 ml of MEM containing L-glutamine (2 mM), penicillin (50 IU/ml) and streptomycin (50 μg/ml) were added. Embryos were incubated at 37.5 °C, humidity 80%, for 2 days, at which point treatments to assay angiogenesis inhibition were started with drugs included into low-melting point agarose for sustained release (Dobson et al., 1990). Sterilized agarose solution was kept at 38 °C; suramin, heparan sulfate HHS-5 or the combination of heparin sodium and hydrocortisone 21-phosphate were added and solutions were poured on 35-mm sterile dishes and placed at 4 °C. Agarose was cut into disks (area: 20 mm^2 and concentration: 2.5%) and peeled away from the plastic surface. Each disk contained 50 μg of suramin, 100 μg of heparan sulfate HHS-5 or 60 μg of heparin sodium and 50 μg of hydrocortisone 21-phosphate. Disks were placed on the CAM of embryos which were returned to the incubator for 2 days. To evaluate the CAM vascular network, a 10% fat emulsion for intravenous use (Lipofundin®) was injected into the chorioallantois and image analysis of the CAM was performed (Tanaka et al, 1986).

Rat cornea assay
The experimental procedure of Proia et al. (1988) was used, with minor modifications. Briefly, both corneas of anesthetized rats were cauterized by pressing the tip of a $AgNO_3/KNO_3$ (1:1, w/w) stick for 30 seconds to the surface at a point approximately 2 mm from the corneoscleral limbus. Suramin, heparan sulfate HHS-5, standard or low molecular weight (LMW) heparin sodium alone or in combination with hydrocortisone 21-phosphate were dissolved in 3.5% methylcellulose. One our after cauterization rats were given two drops of the drug solutions, and the treatment was repeated four times daily for 6 days. At the end of treatment, corneal vessels were visualized by perfusing the upper half of deeply anesthetized rats with a mixture of 10% india ink and 11% gelatin in lactated Ringer's solution. The eyes were then rapidly

cooled using a stream of dichlorodifluoromethane to solidify the gelatin mixture, enucleated and placed in phosphate-buffered neutral formaldehyde for 24 hours. The cornea and the adjacent scleral tissue were cut and magnified images of flat preparations were photographed. Histologic examination of hematoxylin and eosin-stained sections of the cornea adjacent to the lesion was performed.

Results

Suramin and heparan sulfate dose-dependently inhibited the proliferation of BAEC in log-phase of growth (Fig. 1). In addition to this, both drugs reduced the mitogenic effect of bFGF on quiescent cells (Fig. 1).

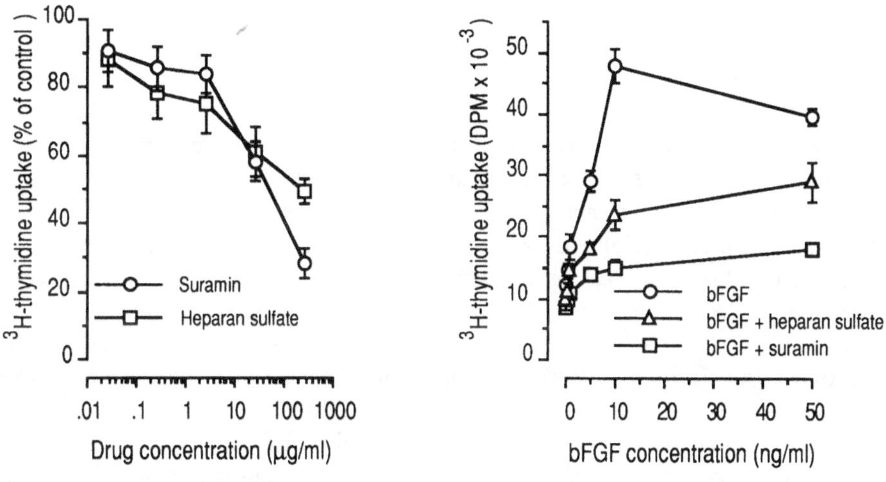

Figure 1. Effect of suramin and heparan sulfate on the proliferation of BAEC in log-phase of growth (left) or stimulated by bFGF (right). The concentration of suramin and heparan sulfate for experiments with bFGF was 300 µg/ml. Each point is the mean of triplicate experiments with SEM shown by vertical bars.

Suramin (50 µg) was a more potent inhibitor of angiogenesis in the CAM than heparin (60 µg) + hydrocortisone 21-phosphate (50 µg) and as active as heparan sulfate HHS-5 (100 µg); the mean avascular area produced by the three treatments was 80, 58 and 74 mm^2, respectively.

In corneas after chemical cauterization there was a marked dilation of the limbal vasculature and a dense brushwork of vessels extended into the cornea towards the lesion; the new vessels were confined to the region between the cautery site and the limbus and formed extensive anastomoses. Histologic examination of cornea revealed infiltration of leukocytes throughout the thickness of the stroma and vascular sprouts projecting from the limbal vasculature could be detected histologically. Suramin strongly inhibited the growth of blood vessels; standard heparin alone did not affect the growth of blood vessels while LMW heparin reduced the amount and lenght of vessel branches. Combination of standard or LMW heparin with hydrocortisone 21-phosphate markedly inhibited the vasculogenesis. Treatment with the heparan sulfate HHS-5 reduced the number of vascular sprouts and few, if any, inflammatory cells could be detected in the corneal stroma.

Discussion

The crucial role of bFGF in vasculogenesis has been demostrated by several authors in different experimental models (for review see Risau, 1990). A recent report has provided further evidence that the bFGF secreted by tumor cells, in which the leader sequence-fused bFGF-gene was transfected, stimulated the development of a vascular network within the tumor transplanted in the mouse; its growth was inhibited by the administration of an immunoneutralizing monoclonal antibody against bFGF (Hori et al, 1991).

The findings of the present study demonstrate that suramin and heparan sulfate inhibit endothelial cell growth and angiogenesis in the CAM and in the rat cornea after chemical cauterization. The results indicate that these polysulfonated compounds are at least as active as heparin and hydrocortisone, a drug combination which has been shown to be effective in angiogenesis inhibition (Harada et al, 1992). The mechanism of the antiangiogenic effect of suramin and heparan sulfate remains to be elucidated; however, due to their polyanionic structure, these drugs may effectively bind polycationic angiogenic factors and prevent their biologic activity. Under this point of view, the results obtained provide further evidence that heparin-binding growth factors, such as bFGF, play a relevant role in several and unrelated models of *in vivo* and *in vitro* vascular growth, including those used in the present study.

In conclusion, suramin and heparin-like molecules are negative regulators of endothelial cell growth and angiogenesis in the CAM and in the rat cornea after chemical injury; their pharmacologic effect may be of value in the treatment of diseases in which persistent and uncontrolled growth of blood vessels play a relevant pathogenetic role.

References

Baldin V, Roman A-M, Bosc-Bierne I, Amalric F, Bouche G (1990) Translocation of bFGF to the nucleus is G_1 phase cell cycle specific in bovine aortic endothelial cells. EMBO J 9:1511-1517

Dobson DE, Kambe A, Block E, Dion T, Lu H, Castellot JJ Jr, Spiegelman BM (1990) 1-Butyryl-glycerol: a novel angiogenesis factor secreted by differentiating adipocytes. Cell 61:223-230

Dugan JD Jr, Lawton MT, Glaser B, Brem H (1991) A new technique for explantation and *in vitro* cultivation of chicken embryos. Anat Rec 229:125-128

Folkman J (1992) Angiogenesis: retrospect and outlook. In: Steiner R, Weisz PB, Langer R (eds) Angiogenesis: key principles, science, technology and medicine. Birkhäuser Verlag, Basel, p 4-13

Harada I, Kikuchi T, Shimomura Y, Yamamoto M, Ohno H, Sato N (1992) The mode of action of anti-angiogenic steroid and heparin. In: Steiner R, Weisz PB, Langer R (eds) Angiogenesis: key principles, science, technology and medicine. Birkhäuser Verlag, Basel, p 445-448

Hori A, Sasada R, Matsutani E, Naito K, Sakura Y, Fujita T, Kozai Y (1991) Suppression of solid tumor growth by immunoneutralizing monoclonal antibody against human basic fibroblast growth factor. Cancer Res 51:6180-6184

La Rocca RV, Stein CA, Danesi R, Myers CE (1990) Suramin, a novel antitumor compound. J Steroid Biochem Mol Biol 37:893-898

Myers C, Cooper M, Stein C, La Rocca RV, McClellan MW, Weiss G, Choyke P, Dawson N, Steinberg S, Uhrich MM, Cassidy J, Kohler DR, Trepel J, Linehan M (1992) Suramin: a novel growth factor antagonist with activity in hormone-refractory metastatic prostate cancer. J Clin Oncol 10:881-889

Proia AD, Chandler DB, Haynes WL, Smith CF, Suvarnamani C, Erkel FH, Klintworth GK (1988) Quantitation of corneal neovascularization using computerized image analysis. Lab Invest 58:473- 479

Risau W (1990) Angiogenic growth factors. Prog Growth Factor Res 2:71-79

Saksela O, Moscatelli D, Sommer A, Rifkin DB (1988) Endothelial cell-derived heparan sulfate binds basic fibroblast growth factor and protects it from proteolitic degradation. J Cell Biol 107:743-751

Tanaka NG, Sakamoto N, Tohgo A, Nishiyama Y, Ogawa H (1986) Inhibitory effects of anti-angiogenic agents on neovascularization and growth of the chorioallantoic membrane (CAM). The possibility of a new CAM assay for angiogenesis inhibition. Exp Pathol 30:143-150

ROLE OF GANGLIOSIDES IN THE MODULATION OF THE ANGIOGENIC RESPONSE

M. Ziche, L. Morbidelli, A. Parenti, G. Alessandri*, F. Ledda and P.M. Gullino*

Department of Pharmacology

University of Florence

Viale Morgagni 65

50134 Firenze

Italy

The unrestrained growth and metastatic potential of a tumor is strictly correlated to its capacity to induce an efficient capillary network in the host tissue (Folkman J and Cotran R 1976, Ziche M and Gullino PM, 1982). Under physiological circumstances the ability of a given tissue to produce a neovascular response is under strict controll, i.e. the balance between factors and events favoring or interfering with neovascular growth is in equilibrium.

In the adult tissue angiogenesis requires two major events: the presence at microvascular level of 'factors' targeting the mobilization and proliferation of the capillary endothelium and the modification of the microenviroment to favour the organization and the morphogenesis of a new microvascular bed (Ziche et al 1982, De Cristian et al 1990).

The work done in these last years in different laboratories has indicated that many

---

*Dept. Biol Sciences, University of Turin, Via Santena 7, 10126 Turin, Italy

NATO ASI Series, Vol. H 75
Cancer Therapy
Edited by N. D'Alessandro, E. Mihich, L. Rausa, H. Tapiero, and T. R. Tritton

angiogenic factors were also normally present in tissues were angiogenesis did not appear suggesting that the angiogenic response could depend on local activation or inactivation of these molecules (Folkman and Klagsburn 1987, Vlodawskyl et al 1991).

Our previous work let us to hypothesize that molecules acting as triggers and/or modulators of the angiogenic response should be present in the tissue being invaded by the newly formed vessels and the objective of our work has been the search of such molecules (Ziche et al 1982, Alessandri G et al 1986, Ziche et al 1989).

To pursue this objective the composition of a tissue was analyzed at the onset of angiogenesis on the assumption that any change occuring at this time could reveal nature and role of molecules involved in the process.

Angiogenesis was reproduced *in vivo* by inducing neovascular growth into the corneal stroma of New Zealand White rabbits (Gullino PM, 1981). Rabbit cornea was utilized as an experimental model because avascular, i.e. the background is zero, transparent i.e. growing capillaries budding from preexixting vessels can be easily monitored with a stereomicroscope and it is bilateral, i.e. control in the same animal is feasible. Capillary growth into the corneal stroma was obtained by delivering in a slow and constant feature appropriate concentration of the angiogenesis inducer incorporated into a slow release polymer (Elvax-40, DuPont) (Langer aand Folkman, 1976). Prostaglandin E1 (PGE1) and recombinant basic fibroblast growth factor (r-b-FGF) were utilized as effective angiogenesis triggers in the rabbit cornea and because quite different in chemical and biological characteristics. After 60-72 hs capillaries were already budding from the preexisting vessels and ready to invade the avascular corneal stroma underlying the implant, but the cornea was not yet colonized by the newly-formed vessels.

When exposed *in vitro* to this corneal fragment capillary endothelial cells improved dramatically their ability to migrate, a prerequisite for new capillary growth

(Alessandri et al 1984). Therefore was reasonable to assume that in the tissue undergoing an angiogenesis process there was an increment in the content of molecules favouring endothelial cell adhesion and movement. The rabbit cornea was removed and analyzed at this early stage and its composition was compared to the contralateral cornea of the same animal bearing a control polymer deprived of angiogenic activity (Ziche et al 1982). The HPTLC analysis of the tissue revealed that under these conditions the cornea accumulated gangliosides (Ziche et al 1989). The total concentration of sialic acid bearing molecules like GM3, GM2 and GD3* was about double that of control, but the increment for each gangliosides varied (GD3>GM2>GM3) .

This observation prompted us to verify the hypothesis wether the local enrichment of the avascular cornea with selective gangliosides could influence the evolution of the angiogenesis process. The enrichment of rabbit cornea with gangliosides with different sialic acid content never induced a neovascularization process per se. When subliminal concentration of angiogenesis factors were released into the corneal stroma enriched with selective gangliosides (GM1, GD3 and GT1b), the newly formed vascular network was strongly sustained during its onset and morphogenesis. When the corneal content of GM3 was selectively increased angiogenesis was on the contrary repressed (Ziche et al 1992).

Growth and mobilization of microvascular endothelium *in vitro* were also evaluated in presence of gangliosides mixture. Both parameters appeared enhanced when GD3 concentration was prevalent but GM3 prevalence resulted in a reduction of both growth and motility (Alessandri et al 1992). The experiment *in vitro* confirmed that the accumulation of gangliosides observed *in vivo* , when angiogenic stimuli were applied to the cornea, was indeed a condition which favoured angiogenesis. However the prevalence of one ganglioside over the other had a modulating effect on growth and motility of microvascular endothelium.

---

* Abbreviations according to I. Svenneeerholm , J Lipid Res 5: 145-155 (1964).

CONCLUSIONS

Shedding of gangliosides is a common event for many neoplastic cell populations and this alters the ganglioside composition of the microenviroment. This change may favour or repress angiogenesis which is a necessary prerequisite for a neoplastic microembolous to produce a metastasis. The data reported indicate that angiogenesis may be favoured or repressed by molecules normally present in the interstitial compartment and acting antagonistically according to their local concentration. Changes in the relative proportion of molecules normally present in adult tissue, like PGE1, FGF GM3 and GD3 can be sufficient to modulate or even block angiogenesis. The finding of these molecules and the assessement of these interactions offer a new and promising strategy for the development of a new therapeutical approach to "angiogenesis-dependent diseases."

REFERENCES

Alessandri G, Raju KS, Gullino PM (1986) Interaction of gangliosides with fibronectin in the mobilization of capillary endothelium. Invasion Metastasis 6:145-165.

Alessandri G, De Cristian G, Ziche M Cappa APM, Gullino PM (1992) Growth and motility of microvascular endothelium are modulated by the relative concentration of gangliosides in the medium. J Cell Physiol 151:23-28

Alessandri G, Raju KS, Gullino PM (1984) Angiogenesis *in vivo* and selective mobilization of capillary endothelium *in vitro* by heparin-copper complex. Microcirculation, Endothelium adn Lymphatics 1:329-346

De Cristian G, Morbidelli L, Alessandri G, Ziche M, Cappa APM, Gullino PM (1990) Synergism between gangliosides and basic fibroblast growth factor in favouring survival, growth and motility of capillary endothelium. J Cell Physiol 144:505-510

Folkman J, Cotran R (1976) Relation of vascular proliferation to tumor growth. Int Rev Exp Pathol 16:207.

Folkman J, Klagsburn M (1987) Angiogenic factors. Science 235: 442-447

Gullino PM (1981) Angiogenesis factors "in: Baserga R (ed) Handbook Exp Pharmacology" Springer Verlag, New York, 57: 427-449

Langer R, and Folkman J (1976) Polymers for the sustained release of proteins and other macromolecules. Nature 263:797-800

Vlodawskyl, Fuks Z, Ishai-Michael R, Baskin P, Levi E, Korner G, Bar-Shavit R, Klagsburn M (1991) Extracellular-matrix-resident fibroblastic growth factor: implication for control of angiogenesis. J Cell Biochem 45: 167-176

Ziche M, Jones J, Gullino PM (1982) Role of prostaglandin E1 and copper in angiogenesis. J Natl Cancer Inst 69:475-482

Ziche M, and Gullino PM (1982) Angiogenesis and neoplastic progression *in vitro*. J Natl Cancer Inst 69: 483-487

Ziche M, Alessandri G, Gullino PM (1989) Gangliosides promote the angiogenic response. Lab Inv 61:629-634

Ziche M, Morbidelli L, Alessandri G, Gullino PM (1992) Angiogenesis can be stimulated *in vivo* by a change in the GM3:GD3 ganglioside ratio. Lab Inv *in press*

SUBJECT INDEX

A

B

C

D

E

F

G

H

I

J

K

L

M

N

O

P

R

S

T

U

V

X

NATO ASI Series H

Vol. 21: Cellular and Molecular Basis of Synaptic Transmission. Edited by H. Zimmermann. 547 pages. 1988.

Vol. 22: Neural Development and Regeneration. Cellular and Molecular Aspects. Edited by A. Gorio, J. R. Perez-Polo, J. de Vellis, and B. Haber. 711 pages. 1988.

Vol. 23: The Semiotics of Cellular Communication in the Immune System. Edited by E.E. Sercarz, F. Celada, N.A. Mitchison, and T. Tada. 326 pages. 1988.

Vol. 24: Bacteria, Complement and the Phagocytic Cell. Edited by F. C. Cabello und C. Pruzzo. 372 pages. 1988.

Vol. 25: Nicotinic Acetylcholine Receptors in the Nervous System. Edited by F. Clementi, C. Gotti, and E. Sher. 424 pages. 1988.

Vol. 26: Cell to Cell Signals in Mammalian Development. Edited by S.W de Laat, J.G. Bluemink, and C.L. Mummery. 322 pages. 1989.

Vol. 27: Phytotoxins and Plant Pathogenesis. Edited by A. Graniti, R. D. Durbin, and A. Ballio. 508 pages. 1989.

Vol. 28: Vascular Wilt Diseases of Plants. Basic Studies and Control. Edited by E. C. Tjamos and C. H. Beckman. 590 pages. 1989.

Vol. 29: Receptors, Membrane Transport and Signal Transduction. Edited by A. E. Evangelopoulos, J. P. Changeux, L. Packer, T. G. Sotiroudis, and K.W.A. Wirtz. 387 pages. 1989.

Vol. 30: Effects of Mineral Dusts on Cells. Edited by B.T. Mossman and R.O. Begin. 470 pages. 1989.

Vol. 31: Neurobiology of the Inner Retina. Edited by R. Weiler and N.N. Osborne. 529 pages. 1989.

Vol. 32: Molecular Biology of Neuroreceptors and Ion Channels. Edited by A. Maelicke. 675 pages. 1989.

Vol. 33: Regulatory Mechanisms of Neuron to Vessel Communication in Brain. Edited by F. Battaini, S. Govoni, M.S. Magnoni, and M. Trabucchi. 416 pages. 1989.

Vol. 34: Vectors asTools for the Study of Normal and Abnormal Growth and Differentiation. Edited by H. Lother, R. Dernick, and W. Ostertag. 477 pages. 1989.

Vol. 35: Cell Separation in Plants: Physiology, Biochemistry and Molecular Biology. Edited by D. J. Osborne and M. B. Jackson. 449 pages. 1989.

Vol. 36: Signal Molecules in Plants and Plant-Microbe Interactions. Edited by B.J.J. Lugtenberg. 425 pages. 1989.

Vol. 37: Tin-Based Antitumour Drugs. Edited by M. Gielen. 226 pages. 1990.

NATO ASI Series H

Vol. 38: The Molecular Biology of Autoimmune Disease.
Edited by A.G. Demaine, J-P. Banga, and A.M. McGregor. 404 pages. 1990.

Vol. 39: Chemosensory Information Processing.
Edited by D. Schild. 403 pages. 1990.

Vol. 40: Dynamics and Biogenesis of Membranes.
Edited by J. A. F. Op den Kamp. 367 pages. 1990.

Vol. 41: Recognition and Response in Plant-Virus Interactions.
Edited by R. S. S. Fraser. 467 pages. 1990.

Vol. 42: Biomechanics of Active Movement and Deformation of Cells.
Edited by N. Akkas. 524 pages. 1990.

Vol. 43: Cellular and Molecular Biology of Myelination.
Edited by G. Jeserich, H. H. Althaus, and T. V. Waehneldt. 565 pages. 1990.

Vol. 44: Activation and Desensitization of Transducing Pathways.
Edited by T. M. Konijn, M. D. Houslay, and P. J. M. Van Haastert. 336 pages. 1990.

Vol. 45: Mechanism of Fertilization: Plants to Humans.
Edited by B. Dale. 710 pages. 1990.

Vol .46: Parallels in Cell to Cell Junctions in Plants and Animals.
Edited by A. W Robards, W. J . Lucas, J . D. Pitts, H . J . Jongsma, and D. C. Spray. 296 pages. 1990.

Vol. 47: Signal Perception and Transduction in Higher Plants.
Edited by R. Ranjeva and A. M. Boudet. 357 pages. 1990.

Vol. 48: Calcium Transport and Intracellular Calcium Homeostasis.
Edited by D. Pansu and F. Bronner. 456 pages. 1990.

Vol. 49: Post-Transcriptional Control of Gene Expression.
Edited by J. E. G. McCarthy and M. F. Tuite. 671 pages. 1990.

Vol. 50: Phytochrome Properties and Biological Action.
Edited by B. Thomas and C. B. Johnson. 337 pages. 1991.

Vol. 51: Cell to Cell Signals in Plants and Animals.
Edited by V. Neuhoff and J. Friend. 404 pages. 1991.

Vol. 52: Biological Signal Transduction.
Edited by E. M . Ross and K . W. A. Wirtz. 560 pages. 1991.

Vol. 53: Fungal Cell Wall and Immune Response.
Edited by J. P. Latge and D. Boucias. 472 pages. 1991.

Vol. 54: The Early Effects of Radiation on DNA.
Edited by E. M. Fielden and P. O'Neill. 448 pages. 1991.

Vol. 55: The Translational Apparatus of Photosynthetic Organelles.
Edited by R. Mache, E. Stutz, and A. R. Subramanian. 260 pages. 1991.

NATO ASI Series H

Vol. 56: Cellular Regulation by Protein Phosphorylation.
Edited by L. M. G. Heilmeyer, Jr. 520 pages. 1991.

Vol. 57: Molecular Techniques in Taxonomy.
Edited by G . M . Hewitt, A. W. B. Johnston, and J. P. W. Young .
420 pages. 1991.

Vol. 58: Neurocytochemical Methods.
Edited by A. Calas and D. Eugene. 352 pages. 1991.

Vol. 59: Molecular Evolution of the Major Histocompatibility Complex.
Edited by J. Klein and D. Klein. 522 pages. 1991.

Vol. 60: Intracellular Regulation of Ion Channels.
Edited by M. Morad and Z. Agus. 261 pages. 1992.

Vol. 61: Prader-Willi Syndrome and Other Chromosome 15q Deletion Disorders.
Edited by S. B. Cassidy. 277 pages. 1992.

Vol. 62: Endocytosis. From Cell Biology to Health, Disease and Therapie.
Edited by P. J. Courtoy. 547 pages. 1992.

Vol. 63: Dynamics of Membrane Assembly.
Edited by J. A. F. Op den Kamp. 402 pages. 1992.

Vol. 64: Mechanics of Swelling. From Clays to Living Cells and Tissues.
Edited by T. K. Karalis. 802 pages. 1992.

Vol. 65: Bacteriocins, Microcins and Lantibiotics.
Edited by R. James, C. Lazdunski, and F. Pattus. 530 pages. 1992.

Vol. 66: Theoretical and Experimental Insights into Immunology.
Edited by A. S. Perelson and G. Weisbuch. 497 pages. 1992.

Vol. 67: Flow Cytometry. New Developments.
Edited by A. Jacquemin-Sablon. 1993.

Vol. 68: Biomarkers. Research and Application in the Assessment of
Environmental Health. Edited by D. B. Peakall and L. R. Shugart.
138 pages. 1993.

Vol. 69: Molecular Biology and its Application to Medical Mycology.
Edited by B. Maresca, G. S. Kobayashi, and H. Yamaguchi. 271 pages.
1993.

Vol. 70: Phospholipids and Signal Transmission.
Edited by R. Massarelli, L. A. Horrocks, J. N. Kanfer, and K. Löffelholz.
448 pages. 1993.

Vol. 71: Protein Synthesis and Targeting in Yeast.
Edited by A. J. P. Brown, M. F. Tuite, and J. E. G. McCarthy. 425 pages.
1993.

Vol. 72: Chromosome Segregation and Aneuploidy.
Edited by B. K. Vig. 425 pages. 1993.

NATO ASI Series H

Vol. 73: **Human Apolipoprotein Mutants III. In Diagnosis and Treatment.** Edited by C. R. Sirtori, G. Franceschini, B. H. Brewer Jr. 302 pages. 1993.

Vol. 74: **Molecular Mechanisms of Membrane Traffic.** Edited by D. J. Morré, K. E. Howell, and J. J. M. Bergeron. 429 pages. 1993.

Vol. 75: **Cancer Therapy. Differentiation, Immunomodulation and Angiogenesis.** Edited by N. D'Alessandro, E. Mihich, L. Rausa, H. Tapiero, and T. R. Tritton. 299 pages. 1993.

Vol. 76: **Tyrosine Phosphorylation/Dephosphorylation and Downstream Signalling.** Edited by L. M. G. Heilmeyer Jr. 388 pages. 1993.

www.ingramcontent.com/pod-product-compliance
Lightning Source LLC
Chambersburg PA
CBHW072243240926
51846CB00012B/154
* 9 7 8 3 6 4 2 8 4 6 1 5 1 *